A Practical Approach to Dynamical Systems for Engineers

Woodhead Publishing Series in Mechanical Engineering

A Practical Approach to Dynamical Systems for Engineers

PATRICIA MELLODGE

ELSEVIER

Amsterdam • Boston • Cambridge • Heidelberg
London • New York • Oxford • Paris • San Diego
San Francisco • Singapore • Sydney • Tokyo
Woodhead Publishing is an imprint of Elsevier

WP
WOODHEAD PUBLISHING

Woodhead Publishing is an imprint of Elsevier
80 High Street, Sawston, Cambridge, CB22 3HJ, UK
225 Wyman Street, Waltham, MA 02451, USA
Langford Lane, Kidlington, OX5 1GB, UK

ISBN: 978-0-08-100202-5 (print)
ISBN: 978-0-08-100224-7 (online)

British Library Cataloguing-in-Publication Data
A catalogue record for this book is available from the British Library

Library of Congress Cataloging-in-Publication Data
A catalog record for this book is available from the Library of Congress

CONTENTS

LIST OF FIGURES

LIST OF TABLES

ABOUT THE AUTHOR

Patricia Mellodge is an associate professor of electrical and computer engineering at the University of Hartford. She received a BS in electrical engineering from the University of Rhode Island. Her graduate work was completed at Virginia Tech where she received an MS in mathematics and an MS and PhD in electrical engineering.

Since 2007, Mellodge has been part of the Samuel I. Ward Department of Electrical and Computer Engineering at the University of Hartford. With a focus in control systems and automation, she teaches courses in both the engineering and engineering technology programs. She has worked on projects involving control system design, microwave process design and characterization, and modeling of the human balance system. Prior to this appointment, Mellodge studied at Virginia Tech under Pushkin Kachroo with a focus on dynamical system analysis and control theory. Her dissertation topic involved control system abstraction and its application to mobile robotic control. Also during this time, Mellodge worked in the microwave processing lab in the Department of Materials Science and Engineering on projects involving hardware design for waste remediation and microwave/material interaction.

PREFACE

Dynamical systems are an important topic in engineering. Applications are prevalent within mechanical, electrical, and biomedical engineering and can be found within robotic, automotive, aerospace, and human systems, among others. The purpose of this book is to present dynamical systems topics in a way that is relevant for practicing engineers. It is intended for engineers who need to understand both the background theory *and* how to apply it. As such, there is an attempt to create a bridge between the theory and the application. Every abstract concept is discussed in depth, described in a readable and down-to-earth manner, and illustrated using practical examples. The intended audience is engineers who are working in industry, graduate students who are taking courses or doing research related to dynamical systems, and undergraduate students who are taking courses in control systems. This is not a textbook, and there are no end-of-chapter problems. Rather, it should be considered an application guide for those in the trenches of working with and learning about dynamical systems.

It is assumed that readers have a solid mathematical foundation in calculus, differential equations, and matrix theory. In presenting the material, the emphasis is on applying the theory, so there are relatively few theorems and no proofs. However, there is a *lot* of mathematics. Much mathematical detail is given that is missing from other texts on these topics. The reason for this level of detail is to help readers understand the complete application in real-world systems. The focus is on depth and not breadth. In covering the selected topics at this level of detail, unfortunately, the number of topics had to be limited. As such, this is not a complete and comprehensive presentation of all topics in dynamical systems. However, this book attempts to cover many relevant topics that an engineer in the field would encounter and provide a foundational understanding for further study.

It is also assumed that the reader has some understanding of MATLAB and Simulink, although expertise is not required. Many of the concepts are demonstrated using real-world examples in MATLAB or Simulink. For the earlier chapters, the MATLAB code is explained line by line to show how various concepts are implemented. These explanations are gradually decreased throughout the book.

The book transitions from topics commonly found at the undergraduate level in engineering, to those covered in graduate courses, to those that engineers may never see in a course. As such, the coverage changes in its approach and assumptions about what the reader knows. The layout of the topics is as follows. Chapter 1 introduces dynamical systems, provides motivation for why it's important to study them, and discusses different types of systems. This material should be familiar from undergraduate engineering courses in linear systems and control theory. There is high-level discussion of this material, but the mathematics starts early with definitions of the different classes of systems.

Chapter 2 discusses modeling and covers differential and difference equations, transfer functions, state-space models, eigenvalues, eigenvectors, and singular value decomposition. Although many of these topics are familiar from courses in differential equations, control systems, and linear algebra, the emphasis is on putting them in the context of dynamical systems. There is also an emphasis on working through examples in MATLAB and giving details of the implementation, which may not be covered in those courses.

Chapter 3 focuses on solutions of dynamical equations, equilibrium points, and stability. These concepts are often encountered in introductory graduate-level courses in dynamical systems and control theory. Again, the emphasis is not on deriving the results but applying them. As such, several of the relevant theorems are presented and applied.

Chapter 4 discusses nonlinear systems and some rich behavior that is only found in them such as limit cycles, bifurcations, chaos, and linearization. These are topics typically found in graduate-level engineering courses. There is a minimal amount of theoretical coverage, and the behaviors are described through examples.

Finally, Chapter 5 introduces Hamiltonian systems, which typically fall in the realm of physicists. However, undamped vibrational systems and their equivalent are an important class of Hamiltonian systems, and there is much rich theory in this area. As with Chapter 4, there is minimal theoretical coverage, and the focus is placed more on the introduction of the concepts through examples.

Many people contributed to the creation of the book. Acknowledgement goes out to colleagues and students at the University of Hartford, particularly Lee Townsend and Iman Salehi for their supportive ideas and engaging discussion; colleagues in the van Rooy Center for Complexity and Conflict Analysis for the productive biweekly meetings; Harriet Clayton and Glyn Jones at Elsevier for the support and feedback; and Joe Romagnano for his editorial skills.

Patricia Mellodge
West Hartford, CT

CHAPTER 1

Introduction: What Is a Dynamical System?

1.1 OVERVIEW

Dynamical systems are all around us: from a car traveling down the road to the ripples caused by throwing a pebble into a pond to a clock pendulum swinging back and forth. But what *is* a dynamical system? First, let us explore the words that make up the phrase "dynamical system." The *Merriam-Webster Dictionary* gives these definitions:

> **Dynamic** (adjective): always active or changing; having or showing a lot of energy; of or relating to energy, motion, or physical force
>
> **System** (noun): a group of related parts that move or work together

The term "dynamic" gives us the idea of change or motion and can deal more specifically with physical phenomena. Mathematically, when one thinks of change, derivatives should spring to mind; indeed, these are a key component of how we model dynamical systems.

The definition of "system" involves groups of related parts working together. Although this definition works in general, a diagram such as the one shown in Figure 1.1 best illustrates the concept that we use throughout the book. Simply stated, a **system** is an entity that has an input and an output. A system receives an input and produces an output based on the input and its state.

When the two words are put together to form "dynamical system," things get interesting. A **dynamical system** is one in which inputs, outputs, and even the system characteristics themselves can change with time. The relationship between input and output can be modeled mathematically using various techniques, such as differential and difference equations,

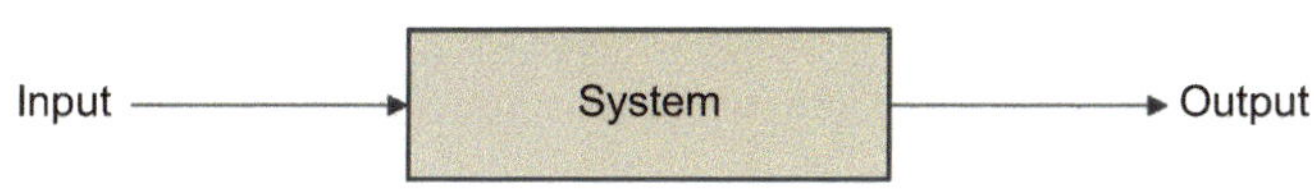

Figure 1.1 Block diagram representation of a system.

A Practical Approach to Dynamical Systems for Engineers
ISBN 978-0-08-100202-5
http://dx.doi.org/10.1016/B978-0-08-100202-5.00001-2

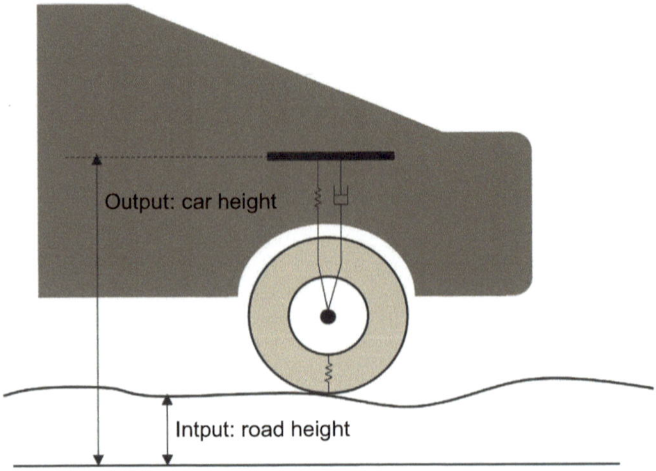

Figure 1.2 A car's suspension system with the road height as the input and the car's height as the output.

transfer functions, and state space equations. Throughout this book, we focus on two objectives: (1) investigate various techniques and analysis methods for dynamical systems and (2) apply these methods to examples of real-world systems and show how they can be used in practice. As we will see, the study of dynamical systems can reveal some very interesting behaviors, some desirable and some not.

What about a car driving down a road makes it a dynamical system? One example is its suspension system, represented in Figure 1.2. The input to this system is the road, which has a varying height as the car drives down it. The output might be the ride height of the car itself. The system consists of the various components linking the road to the passenger's body: the tires, control arm, shock absorber, chassis, and so on. When you hit a bump in the road, you experience a dynamical response: perhaps your body is jarred by the sudden change in wheel height, or perhaps you barely notice it. The response depends on the characteristics of the car's suspension, which can be modeled as a mass-spring-damper system.

1.1.1 Why Do We Study Dynamic Systems?

The main reasons are (1) to **predict** system behavior and (2) to **control** system behavior.

An example of a system we study for prediction is the weather. Scientists, meteorologists in particular, have spent many years and many computing hours creating models used to predict the weather. We rely on these predictions to plan our daily activities, from scheduling an event to deciding what to wear. As computing power advanced and more atmospheric data became available, the models used to predict the weather increased in sophistication and accuracy.

The above example of the car suspension is another case of a systemic behavior we wish to predict. Knowing the effects of variables such as spring stiffness, fluid viscosity in shock absorbers, tire pressure, and the many other parameters that go into the system allow the designers to predict and therefore design a system to have the desired characteristics. Certainly, someone buying a sports car would be disappointed if the vehicle handled like a school bus!

There are many examples of systems whose behavior we wish to control. The previous example of the car suspension is also one of them. Many modern cars now have adaptive suspension systems in which the vehicle performance through turns and over bumps in the road is monitored. Based on this performance, the system is designed to adjust itself by changing the damping coefficient of the shock absorbers.

Another example of controlling a system from everyday life is a thermostat like the one shown in Figure 1.3. A thermostat allows us to set a comfortable temperature for the room we are in, and this temperature is maintained automatically. The temperature of the room depends on many parameters, including heat sources within the room (e.g., heaters, lights,

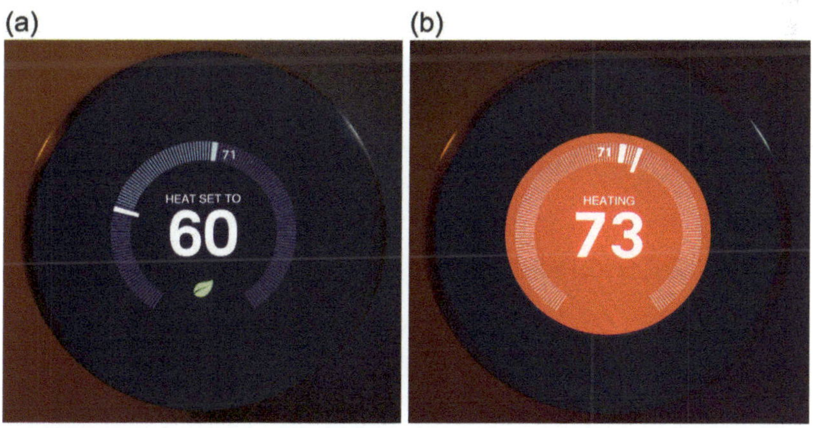

Figure 1.3 A thermostat automatically controls the temperature of a space. In (a), the temperature is above the set point, and heating does not occur. In (b) the desired temperature is above the room temperature, and the system is heating.

people); the amount of insulation in the walls, doors, and windows; and the temperature outside the room. Designers wishing to optimize thermostat performance need to have an understanding of the temperature dynamics in the room. With this knowledge, the designer can choose the controller parameters to get the best performance from the system.

The analysis strategy for a system depends on the objective. If one is trying to predict system behavior, then often the strategy is to model as much of the characteristics as possible. In the case of weather, an extraordinary amount of data is gathered for use in computer models. This data is taken from many different sensors to measure a number of conditions (e.g., temperature, humidity, barometric pressure) in many different locations. Despite sophisticated modeling techniques, one aspect of weather systems that makes prediction so difficult is its inherently chaotic behavior. This is exemplified by the thought experiment known as the butterfly effect: a butterfly flapping its wings in one part of the world can cause a hurricane somewhere else weeks later. We will discuss chaos later in Chapter 4.

If a designer is trying to analyze a system to develop a control algorithm, the technique may not be to model as much as possible about the system. In many cases, it is not practical and simply not necessary to obtain so much information. For the example of a thermostat, a comprehensive model of the room would need the temperature distribution in the space. In other words, we would need to measure the temperature at every point in the room. However, this kind of measurement is not practical because we cannot possibly fit that many sensors in the room. It is more likely that we can get away with one, two, or four sensors by making assumptions about the temperature distribution in the room. Furthermore, the control designer often relies on a relatively simple model and focuses on a sophisticated control algorithm to compensate for modeling errors and uncertainty.

1.2 TYPES OF SYSTEMS

Dynamical systems can be categorized by their characteristics. In this section, we discuss several of these categories and give examples of each type. Understanding the type of system you are dealing with is important because it will drive which techniques are available for you to use. Serious problems can occur if analysis and design methods are used on the wrong type of system!

1.2.1 Continuous versus Discrete

Consider the above example of a car suspension system. A plot of the car's ride height versus time after going over a bump may look something like the curve shown in Figure 1.4. At every instant of time, we can get a value for the car's height. A fraction of an instant later, we can get another value. In fact, we can get a different value at every instant in time. This is an example of a **continuous-time system**.

In contrast, consider money in a bank account. Assume an account is earning 5% interest and there is an initial deposit of $10,000. Figure 1.5 shows how the money would grow if no deposits or withdrawals are made and the bank applied interest monthly. Now similar to the car suspension example, we can check the account balance at any time and get a value. However in this case, changes only occur once every month. Because the in-between values are known to be held constant, we can represent the balance as shown in Figure 1.6.

The formula for the amount of money in the account is

$$A_n = A_0\left(1 + \frac{r}{12}\right)^n \tag{1.1}$$

where r is the annual interest rate, n is the number of months, A_0 is the initial deposit, and A_n is the amount of money in the account in month n. This variable n represents a discrete interval of time.

The bank account is an example of a **discrete-time system**. In a discrete-time system, the values are only recorded at particular time instants. The in-between values do not matter or do not change (as in the bank account), or we do not have access to them. Sampling is the process by

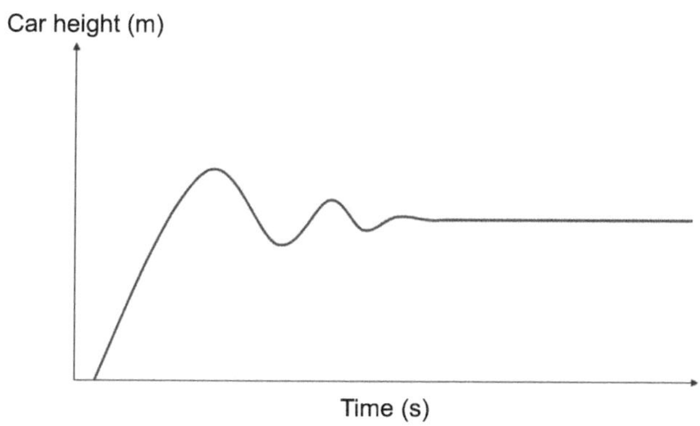

Figure 1.4 The car's height is an example of a continuous-time signal.

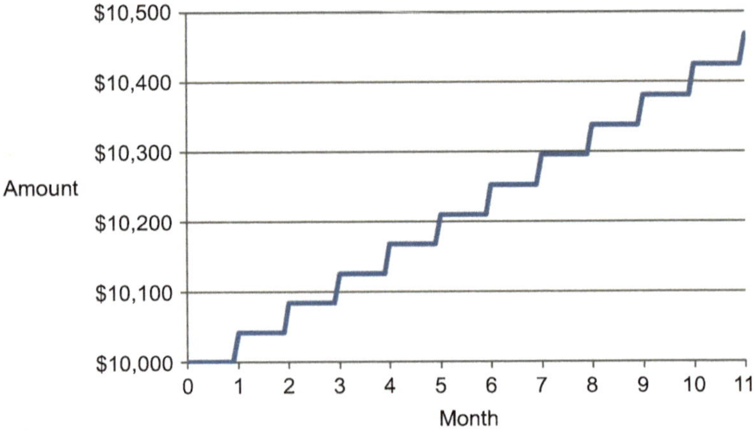

Figure 1.5 An account balance with interest applied monthly represented by a continuous-time signal.

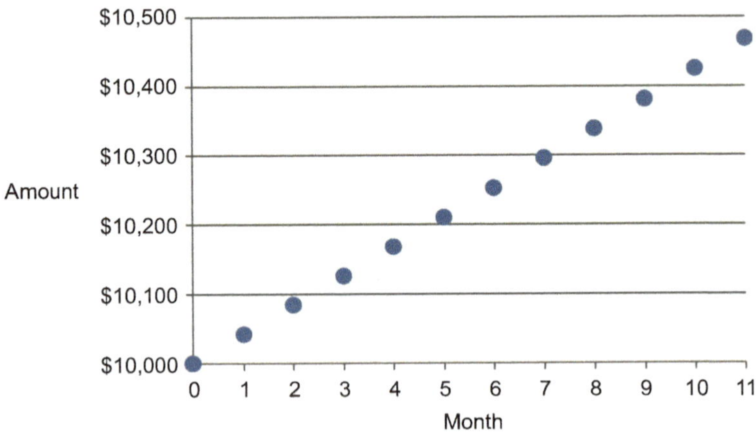

Figure 1.6 The same account balance represented as a discrete-time signal.

which a continuous-time system is converted to discrete time. In this situation, the sampling rate is very important because if the samples are not taken often enough, important information in the signal can be lost.[1] On the other hand, there is always an upper limit on the sampling rate caused by hardware processing limitations, and faster sampling is not always the key to improving system performance.

[1] This issue is addressed by the Nyquist sampling theorem, which states that if the highest frequency in a signal is f_{max}, then the signal can be completely reconstructed from a sampled version if the minimum sampling rate is $2f_{max}$.

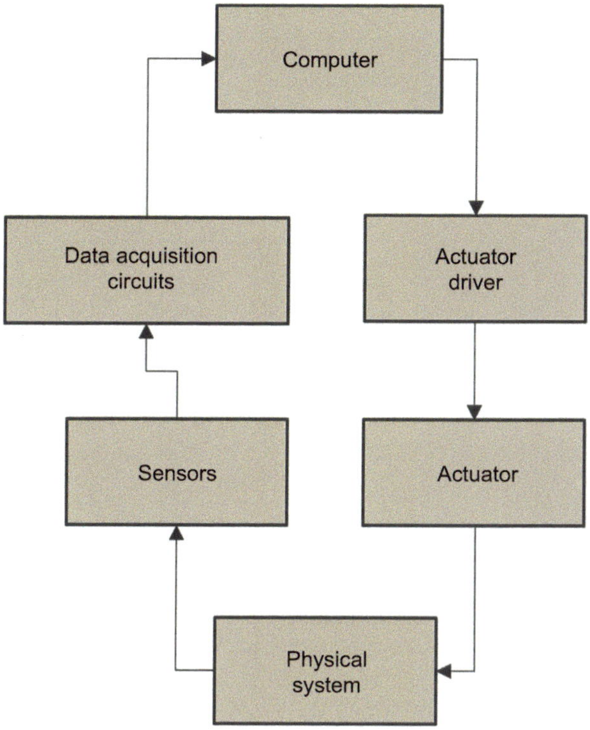

Figure 1.7 A typical computer-controlled system.

Real physical systems, such as the car suspension system, are mostly analog in nature and therefore continuous-time systems. So why study discrete-time systems? The major reason is that many of these systems are computer controlled, and computers are discrete-time machines that run on a clock. Figure 1.7 shows a typical physical system controlled by a computer. Typically, when a computer sends signals to actuators and receives signals from sensors, some form of analog-to-digital and digital-to-analog conversion must take place.

1.2.2 Linear versus Nonlinear

A **linear** system can be characterized in several different, but related, ways.

- Its dynamics can be represented by a system of linear differential equations (for continuous-time systems) or linear difference equations (for discrete-time systems).
- It has a transfer function.
- It obeys the law of superposition.
- A sinusoidal input produces a sinusoidal output of the same frequency.

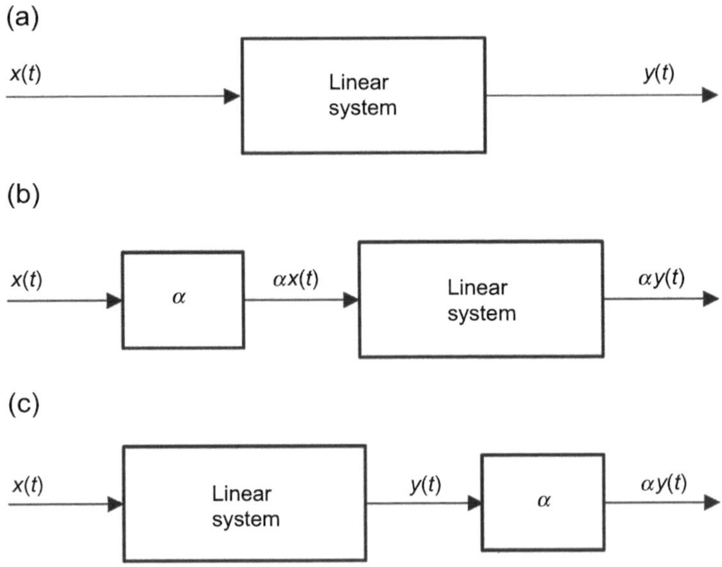

Figure 1.8 The scaling property. For a linear system with input $x(t)$ and output $y(t)$ as in (a), a multiplier block placed before the system (b) will result in the same output as when the multiplier block is after the system (c).

One of the most common ways to test for a system's linearity is by verifying if it follows the law of superposition. Superposition is usually introduced to engineers in a circuit analysis course, but the concept applies to linear systems in general. (Resistive circuits are linear systems after all.) Superposition is composed of two parts, scaling and additivity.

Scaling: If a system's input $x(t)$ results in an output $y(t)$, then an input of $\alpha x(t)$ will result in the output $\alpha y(t)$ for any value α. This characteristic is shown in Figure 1.8.

In other words, if you multiply the system's input by some value, you end up multiplying the output by the same value. Note that this also means zero input should result in zero output.

Additivity: If a system's input $x_1(t)$ results in $y_1(t)$ and $x_2(t)$ results in $y_2(t)$, then if the input is $x_1(t) + x_2(t)$, the output is $y_1(t) + y_2(t)$. A block diagram depicting this characteristic is shown in Figure 1.9.

In other words, in a linear system, it does not matter if you sum the signals before or after the system. The output is the same.

A **nonlinear** system is simply one that is not linear. However, there are several reasons why a system might be nonlinear, and different classes of nonlinearities come about because of different physical reasons.

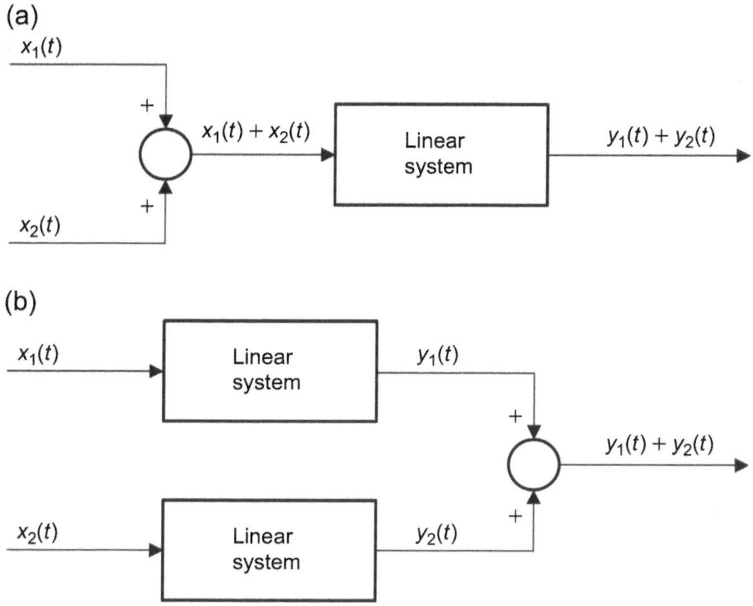

Figure 1.9 In a linear system, signals may be summed together before entering the system (a) or after leaving the system (b). The output signal is the same in both cases.

Let's consider an example of a simple nonlinear system. Suppose a system has input x and output y and they are related by

$$y = mx + b \tag{1.2}$$

where m and b are nonzero constants. This is a linear equation! So how can it be nonlinear? Although it may be a linear equation, the system it represents is nonlinear. We can see this if we look at the scaling property.

According to the scaling property, let $x_0(t)$ be the input and $y_0(t)$ be the corresponding output, as in

$$y_0 = mx_0 + b \tag{1.3}$$

Now let's see what happens when we multiply the input by α and call the output y_1.

$$y_1 = m(ax_0) + b \tag{1.4}$$

And when we multiply the output by α, we get

$$\begin{aligned} ay_0 &= a(mx_0 + b) \\ &= m(ax_0) + ab \end{aligned} \tag{1.5}$$

Because y_1 does not equal αy_0, the system is nonlinear! We can also see this if we put zero into the system, the output is b (not zero).

Let's also take a look at how the system does not obey the additivity property. (This is just an exercise. We have already shown the system is nonlinear because it violates the scaling property.)

Suppose the inputs $x_1(t)$ and $x_2(t)$ result in outputs $y_1(t)$ and $y_2(t)$, respectively. Then

$$y_1 = mx_1 + b \tag{1.6}$$

$$y_2 = mx_2 + b \tag{1.7}$$

Putting the sum of x_1 and x_2 into the system, the output is

$$y_s = m(x_1 + x_2) + b \tag{1.8}$$

But

$$\begin{aligned} y_1 + y_2 &= mx_1 + b + mx_2 + b \\ &= m(x_1 + x_2) + 2b \end{aligned} \tag{1.9}$$

Again we see that the system is nonlinear because $y_s \neq y_1 + y_2$, and the additivity property is not followed.

Here is another example that's practical and not purely mathematical. Consider the amplifier circuit shown in Figure 1.10.

A circuit like this is usually discussed in an introductory circuits or electronics course. It is an inverting amplifier, and the input–output relationship is given by

$$V_{out} = -\frac{R_2}{R_1} V_{in} \tag{1.10}$$

If you double the input voltage, the output voltage is doubled, and the scaling property holds. Also, if you add two inputs, the result is

$$-\frac{R_2}{R_1}(V_1 + V_2) = \left(-\frac{R_2}{R_1} V_1\right) + \left(-\frac{R_2}{R_1} V_2\right) \tag{1.11}$$

and the additivity property holds. Clearly, this must be a linear system.

Mathematically, yes, the system is linear. However, suppose your two input voltages are 10 and 15 V. Will the output be −25 V? No, of course not! The circuit is only being supplied with ±12 V, and the output cannot exceed this limit.

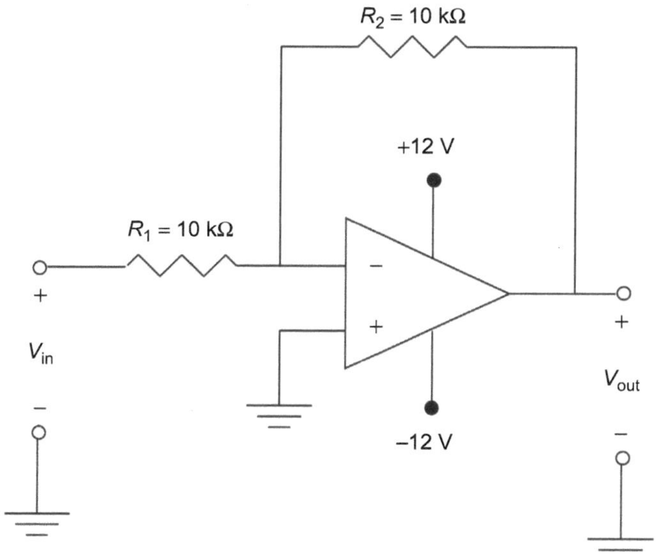

Figure 1.10 Op-amp inverting amplifier circuit.

This op-amp circuit is an example of a system that is modeled linearly and behaves linearly within certain limits (for input voltages less than $\frac{R_1}{R_2}$ 12 V in magnitude). However, when you exceed those limits, the output saturates, and the nonlinearity is seen as a distortion, as shown in Figure 1.11. This limitation is one reason microphones clip and speakers distort sound.

There are many types of nonlinearities, and they can come about from different types of physical limitations such as thresholds, sticking, hysteresis, or other imperfections. Nonlinear systems, and how to deal with them, are discussed in more detail in Chapter 4.

1.2.3 Time-Invariant versus Time-Varying

A **time-invariant** (or **autonomous**) system is one whose behavior does not depend explicitly on time. In other words, given an input, if you run the system now, an hour from now, or next Tuesday, the output will be the same. Mathematically, time invariance means that the "constants" in the system are truly constant and do not vary with time.

As an example, consider a simple first-order model of a car:

$$m\ddot{x}(t) = F(t) - b\dot{x}(t) \tag{1.12}$$

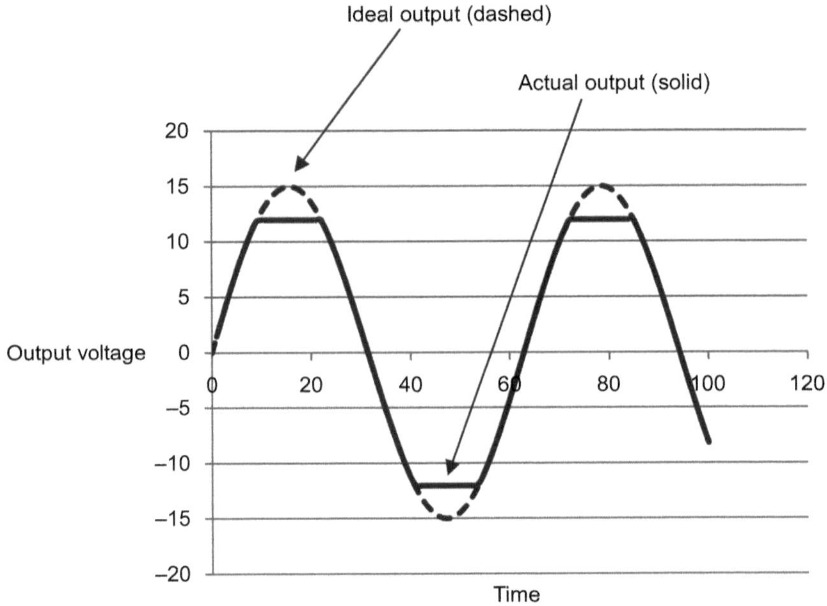

Figure 1.11 The ideal output according to the equation of the inverting amplifier is a sine wave. However, in reality, the actual output voltage is limited to a maximum value, so the sine wave is distorted.

where m is the mass of the car, x is the position, F is the force applied to the wheels by the engine and drivetrain, and b is the coefficient of friction. We model this system as one that is time-invariant because we assume the mass and friction to be constant.

In reality, the car is a **time-varying** (or **nonautonomous**) system. As you drive, the car consumes fuel, and the mass decreases. However, this change is small and occurs slowly, so mass is assumed constant. Many systems we take to be time invariant are not because of aging. In our equations, we use parameters such as mass, friction, length, area, resistance, and many others as constants. But they do change as components age. The time-invariance assumption is valid as long as the timeframe over which the parameters change is much longer than the run time for our system. If it takes many years for a resistor value to change, but if we run the circuit for minutes or days, then the time-invariance assumption works. In the bank account example, the interest rate may change, but the system can be considered time-invariant if the interest rate is constant during the window of time we are investigating.

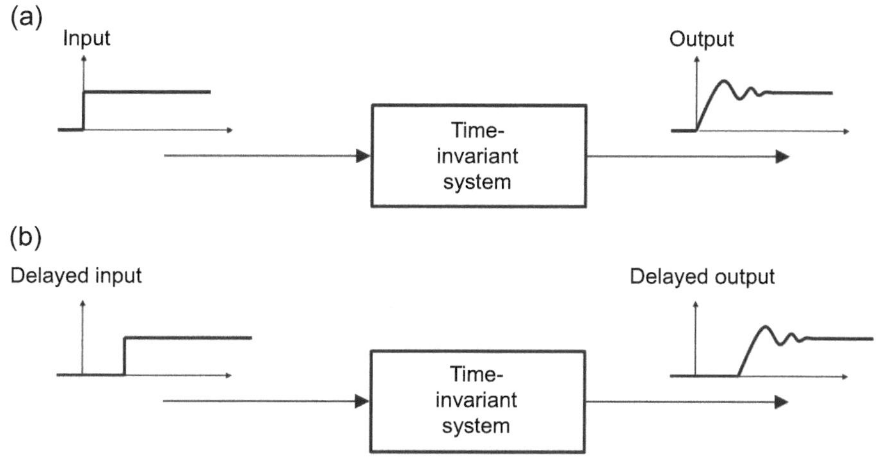

Figure 1.12 If a time-invariant system has an input and output as in (a), then if the input is delayed by a certain amount, the output will be delayed by the same amount (b).

To check to see if a system is time-invariant, one only needs to delay the input and observe the output. Assuming that an input of $x(t)$ produces an output $y(t)$, then the system is time invariant if the input $x(t-\tau)$ gives an output of $y(t-\tau)$ for any value of τ. This relationship is shown in Figure 1.12.

1.2.4 Memory versus Memoryless

A system has **memory** if its output depends at all on past inputs. A system is **memoryless** if its output depends only on the current input. This concept can be seen by the following example.

Consider a spring, with the relationship between position and force given by

$$F = kx \tag{1.13}$$

where k is the spring constant. If you want to know the position of the spring given its force, you simply divide by k. On the other hand, suppose you have a system with friction where

$$F = b\dot{x} \tag{1.14}$$

where b is the coefficient of friction, then given the force, you need to integrate to get position

$$x(t) = \frac{1}{b}\int_{-\infty}^{t} F(\tau)d\tau \tag{1.15}$$

In this case, the force needs to be known for all time up the current time to determine the position.

1.2.5 Causal versus Noncausal

A **causal** system is one whose output depends only on the present and the past inputs. A **noncausal** system's output depends on the future inputs. In a sense, a noncausal system is just the opposite of one that has memory.

How can a real-world system be noncausal? It cannot because real systems cannot react to the future. But noncausal systems have important real-world applications. Consider a song stored in a sound file. Because the entire song is stored, we could process the sound by filtering in a way that has the current notes depend on notes later in the song. This is an example of postprocessing in which noncausal systems may be implemented. Another example of a noncausal system application is image processing. The pixels to the left of the current location can be considered as the "past" and pixels to the right as the "future."

1.2.6 Deterministic versus Stochastic

A **deterministic** system is one in which parameters and inputs are known. Consider the earlier example of the simple first-order model of a car:

$$m\ddot{x}(t) = F(t) - b\dot{x}(t) \tag{1.16}$$

This system is deterministic if we know exactly values for the mass m, friction coefficient b, input force F, and initial conditions of x and $\dot{x}$ and know that there are no other inputs, or disturbances, acting on the system. In this ideal case, we can figure out the system's trajectory (the signals $x(t)$ and $\dot{x}(t)$) simply by solving the differential equation (either analytically or numerically).

Unfortunately, real-world systems usually have some uncertainty in them, and figuring out the system's trajectory is not so simple. If any one of the parameters or inputs in the equations describing the system is not or cannot be known exactly, then the system is stochastic. Also if a system is affected by noise in any part of it, then the system is stochastic.

Stochastic systems are an active field of study involving what are known as stochastic differential equations. In these sorts of systems, the parameters or inputs to the system are not known exactly, but it is assumed that their probabilistic properties are known. For example, it could be assumed that Gaussian white noise is entering the system through the sensor data.

A more thorough discussion and exploration of stochastic systems are beyond the scope of this book. However for interested readers, good starting points are books by Astrom (1970) and Pugachev and Sinitsyn (2001).

1.3 EXAMPLES OF DYNAMICAL SYSTEMS

So far, several examples were introduced in the context of the different system types, such as the car suspension system, bank account, and op-amp circuit. One focus of this book is to provide real-world examples for each of the concepts discussed. The hope is to show how the theory can be applied to real systems. As a result, many more examples will be covered throughout the book to illustrate the concepts being discussed and then to apply them to real-world systems. Some of these examples will carry through multiple chapters, and many analytical methods will be applied to them. Examples of real-world systems presented include:

- Driving a car, both lateral (steering) and longitudinal (speed) dynamics
- Populations of humans and other living organisms
- The human body, in particular the balance system
- Aircraft engines
- Electromechanical systems such as motor-driven devices
- Translational and rotational mechanical systems such as the mass–spring–damper and pendulum

1.4 A NOTE ON MATLAB AND SIMULINK

A second focus of this book is to illustrate analysis methods using the common software tools of MATLAB and Simulink created by Mathworks, Inc. As with the examples, many concepts will involve MATLAB or Simulink (or both) to apply the methods. These programs were chosen because of their popularity in industry and academia.

MATLAB (which stands for MATrix LABoratory) is a powerful program that allows engineers to perform computations based on matrices using a high-level programming language. With this program, one can perform simulations, apply analytical tools, and display plots of results. MATLAB has many toolboxes, which are available to expand the capabilities into many different areas. Some examples of toolboxes relevant to dynamical systems are the Control Systems Toolbox and System Identification Toolbox.

Simulink is an add-on program to MATLAB that has a drag-and-drop interface that allows users to create and simulate systems visually using block diagrams. It has much of the same capabilities as MATLAB but with a graphical interface, and information can be passed back and forth between the two programs.

For this book, there is the expectation that readers have some familiarity with MATLAB and Simulink. However, expertise is not necessary. Someone who has used these programs should be able to follow the examples and reproduce the results, along the way learning new elements of the software and new strategies for using them as analytical tools. Expertise comes with studying examples and practice with creating code. Readers who have not used MATLAB or Simulink should also be able to follow the examples and understand the results being obtained. If further help is needed, several good books are available as MATLAB and Simulink references, including those by Beucher and Weeks (2008), Moore (2014), and Tyagi (2012).

The MATLAB and Simulink examples show one way to address the problem. As with any programming task, there are multiple ways to accomplish the desired result. The provided code may or may not be the "best" program in terms of execution time, memory use, or efficiency. These characteristics were secondary to clarity and ease of understanding because the main goal is conveying the ideas behind how MATLAB and Simulink are used as a tool to solve problems.

REFERENCES

Astrom, K. (1970). *Introduction to Stochastic Control Theory*. Mineola, NY: Dover Publications.

Beucher, O., & Weeks, M. (2008). *Introduction to MATLAB and Simulink: A Project Approach* (3rd ed.). Hingham, MA: Infinity Science Press.

Moore, H. (2014). *MATLAB for Engineers* (4th ed.). Boston: Prentice Hall.

Pugachev, V., & Sinitsyn, I. N. (2001). *Stochastic Systems: Theory and Applications*. River Edge, NJ: World Scientific Publishing.

Tyagi, A. (2012). *MATLAB and SIMULINK for Engineers*. New Delhi: Oxford University Press.

FURTHER READING

Coiffier, J. (2012). *Fundamentals of Numerical Weather Prediction*. Cambridge, UK: Cambridge University Press.

Fayyad, S. (2012). Constructing control system for active suspension system. *Contemporary Engineering Sciences, 5*(4), 189–200.

Guglielmino, E., Sireteanu, T., Stammers, C. W., Gheorghe, G., & Giuclea, M. (2008). *Semi-active Suspension Control: Improved Vehicle Ride and Road Friendliness*. London: Springer-Verlag London.

Jackson, L. (1990). *Signals, Systems, and Transforms*. Reading, MA: Addison-Wesley.

CHAPTER 2

System Modeling

2.1 INTRODUCTION

What is a system model? At the most basic level, a **model of a system** describes how that system behaves. Such descriptions can be in the form of words, diagrams, or equations depending on what is appropriate for the system being modeled. Although systems can include people or businesses (think of role models or business models), we will be concerned with physical systems and their mathematical representation.

In Chapter 1, we discussed how a system can be represented simply and abstractly by a block diagram such as the one shown in Figure 2.1, in which there is an input to the system and an output from the system. In this chapter, the focus is on what happens inside the block and the mathematical relationship between the input and the output.

System models can take several forms, and some of these forms depend on the type of system. Every system we present can be modeled using either differential equations (for continuous-time systems) or difference equations (for discrete-time systems). There is also a class of system known as **hybrid systems** that have both continuous-time and discrete-time characteristics. Physical hybrid systems are typically modeled using differential equations with discrete transitions. A classic example of a hybrid system is a bouncing ball when there are continuous-time dynamics when the ball is in the air and discrete transitions when it hits the ground. In any case, these differential and difference equations are *time domain* models. In Section 2.2, these equations are discussed in detail. These equations can also be generalized into a standard form known as a state-space model, which is covered in Section 2.4.

Although differential and difference equations can tell us a lot about a system, sometimes they do not provide information about the system in

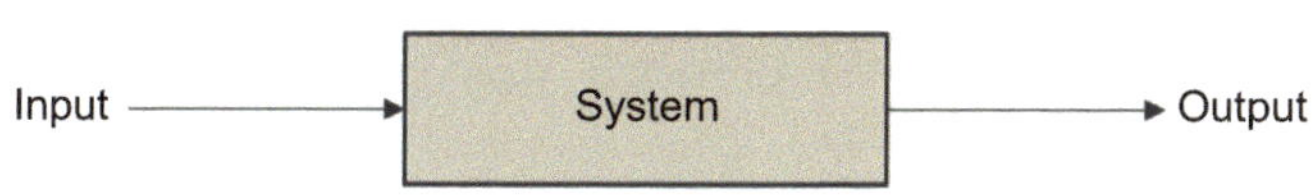

Figure 2.1 Block diagram representation of a system.

A Practical Approach to Dynamical Systems for Engineers
ISBN 978-0-08-100202-5
http://dx.doi.org/10.1016/B978-0-08-100202-5.00002-4

a form in which we can easily identify certain system characteristics. Another way of representing a *linear* system model, which provides an alternative perspective, is in the Laplace domain or z domain using transfer functions. In this form, we can more easily see the frequency response of a system and its stability properties. Transfer functions of both continuous-time and discrete-time systems are discussed in Section 2.3.

Finally, in Section 2.5, the determination of model parameters is discussed. System identification refers to the process of figuring out the constants of a system through experimentation. This process is different from modeling a system from physical principles, as can be done with simpler systems. With system identification, a system is treated more like a black box, and experiments and tests need to be performed to determine the parameters.

It is important to point out that a system's model is only a representation of its behavior and an imperfect one at that. A mathematical model of a real system can never account for everything, and there will always be some source of error to mess up the description it provides. When finding these imperfections, there is always the temptation to "fix" them by modifying the model to better fit the behavior we see. However, this temptation must be kept at bay to prevent our models from becoming too complex. One has to keep in mind the tradeoff between complexity and accuracy and decide when the model is "good enough" for its purpose. Unfortunately, there is no black and white answer to the question of how good is "good enough." The answer is application specific and can only be found through experience.

One example of a system that can have different types of models is vehicle traffic. On the one hand, the traffic can be thought of as fluid flow with varying density as the vehicles get bunched or spread apart. In this scenario, fluid dynamics equations can be used for the model. On the other hand, the model may be formulated to keep track of individual vehicles and their interactions with the other vehicles and the road environment. Then in this scenario, equations of motion for each individual vehicle would be used, and their combination would describe the entire system. These two types of models treat vehicle traffic at different scales. The fluid model is macroscopic, and the individual vehicle models are microscopic. Depending on the application, one level of detail may be more appropriate than the other.

2.2 EQUATIONS OF MOTION

When we refer to **equations of motion**, we are referring to a set of equations that describes how a system changes in time. For the car suspension system previously discussed, these equations describe the up and down motions of the car relative to the road surface as it drives and experiences variations in the road surface. In this example, there truly is motion in the physical sense. Equations of motion, however, are more general. As another example, consider a bank account earning interest. There is no physical motion happening in the account, but the money is in fact "moving" in the sense that it is changing value with time.

In this section, we develop equations of motion for several different systems (car suspension, car kinematics, bank account, and a computer-controlled vehicle) and show how to use MATLAB to represent and solve these equations.

2.2.1 Differential Equations for Continuous-Time Systems

Let us start with the popular example of the car suspension as shown in Figure 1.2. We will first develop the equations of motion from physical principles and then discuss the form and meaning of these equations.

A simplified version of the system is shown in Figure 2.2. This version consists of two masses. The bottom mass m_1 represents the wheel, and it is connected to the ground by a spring (the flexion of the tire). In this simple model, we assume that the tire's damping is negligible (a simplification that leads to a "good enough" model in this case). The top mass m_2 represents the portion of the car body supported by the wheel. The mass m_2 is connected to the wheel mass by a spring and damper (the shocks and linkage connecting them together). The variables x_1 and x_2 denote the height of the wheel and car, respectively, and the arrows indicate the direction of positive values. We assume that $x_1 = 0$ and $x_2 = 0$ are the equilibrium positions of both masses where the springs are not compressed or stretched. We also assume that r is the height of the road above some reference.

To develop the equations of motion for this system, we first draw a free body diagram for each mass and then apply Newton's second law of motion. The free body diagram for m_1 is shown in Figure 2.3. The large arrows indicate the direction of the force being applied to the body caused by the movement of both masses. The dots above x_1 and x_2 indicate derivatives with respect to time.

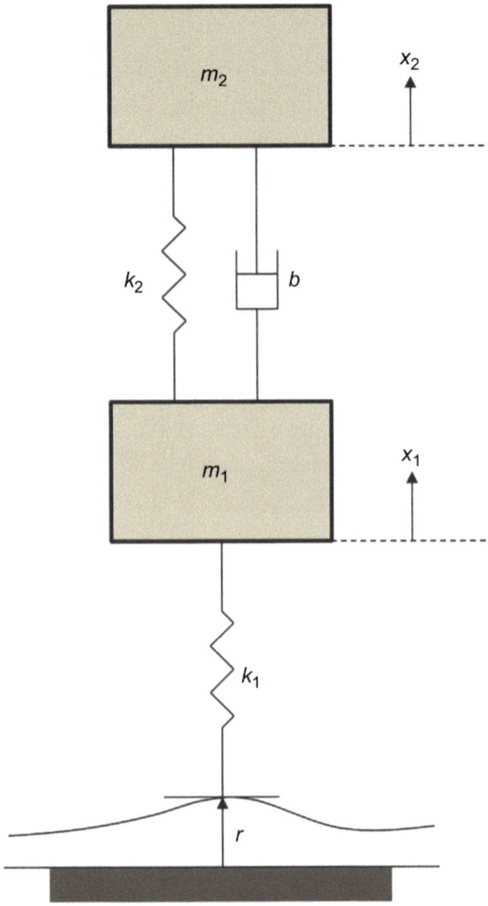

Figure 2.2 Simplified version of the car suspension system.

According to Newton's second law, summing the forces on this body and setting it equal to mass times acceleration gives

$$k_2(x_2 - x_1) + b(\dot{x}_2 - \dot{x}_1) - k_1(x_1 - r) = m_1\ddot{x}_1 \tag{2.1}$$

Note that the effect of gravity can be ignored if we define our displacements, x_1 and x_2, with respect to the equilibrium positions of the springs. Gravity is taken into account when we define the $x_1 = 0$ and $x_2 = 0$ positions by these equilibria because the gravitational force creates an offset that is eliminated by this definition of zero displacement.

We now repeat the process for m_2. First the free body diagram is shown in Figure 2.4.

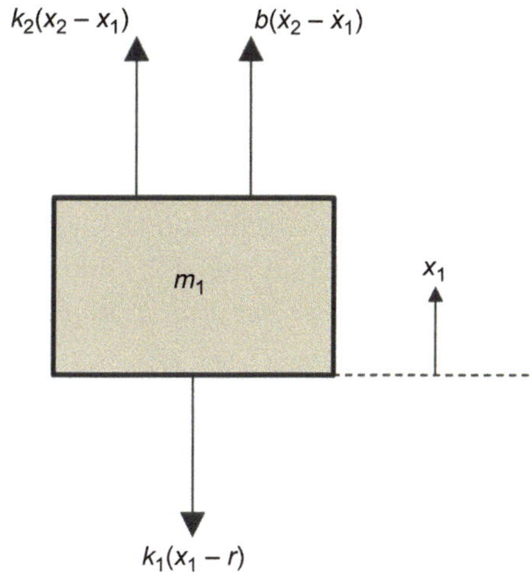

Figure 2.3 The free body diagram for the wheel.

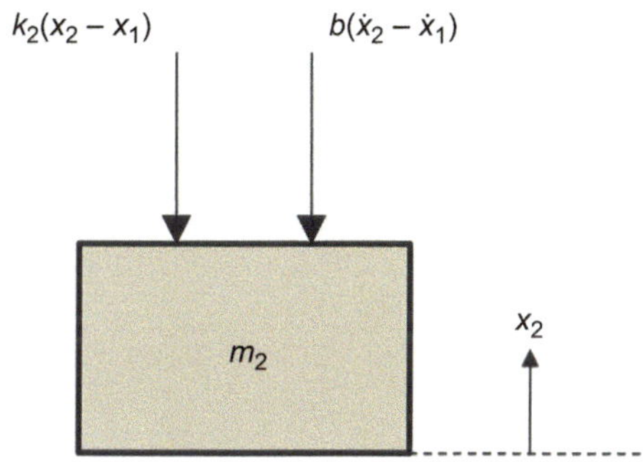

Figure 2.4 Free body diagram for car body.

Then summing the forces on m_2 gives

$$-k_2(x_2 - x_1) - b(\dot{x}_2 - \dot{x}_1) = m_2\ddot{x}_2 \tag{2.2}$$

Finally, the car suspension system dynamics can be represented by the two equations of motion.

$$\begin{aligned} &\ddot{x}_1(t) + \frac{b}{m_1}\dot{x}_1(t) + \frac{k_1 + k_2}{m_1}x_1(t) - \frac{b}{m_1}\dot{x}_2(t) - \frac{k_2}{m_1}x_2(t) = \frac{k_1}{m_1}r(t) \\ &\ddot{x}_2(t) + \frac{b}{m_2}\dot{x}_2(t) + \frac{k_2}{m_2}x_2(t) - \frac{b}{m_2}\dot{x}_1(t) - \frac{k_2}{m_2}x_1(t) = 0 \end{aligned} \tag{2.3}$$

Note that these equations describe a fourth-order system (i.e., there are two states x_1 and x_2, each with a second derivative). The final form in (2.3) is in **monic form**. Also note that the variable t has been added to show explicitly which variables depend on time and which do not. However the (t) is often dropped for brevity. The characteristics of monic form are:

1. State variables (x_1 and x_2) are collected on the left side of the equation.
2. Input variables (r) are collected on the right side of the equation.
3. State variables are collected together and arranged in descending order of the derivatives.
4. The highest derivative term in each equation has a coefficient of 1.

It is useful to put equations in a standard form because it flows easily into the concepts of transfer functions and state-space representations of system. We will see this connection when these concepts are discussed in Sections 2.3 and 2.4.

The model in (2.3) is an **ordinary differential equation** (ODE) because it has one independent variable (t). In contrast, **partial differential equations** (PDEs) have more than one independent variable. Partial differential equations are not discussed in this book, but interested readers may consult Farlow (1993) as a starting point.

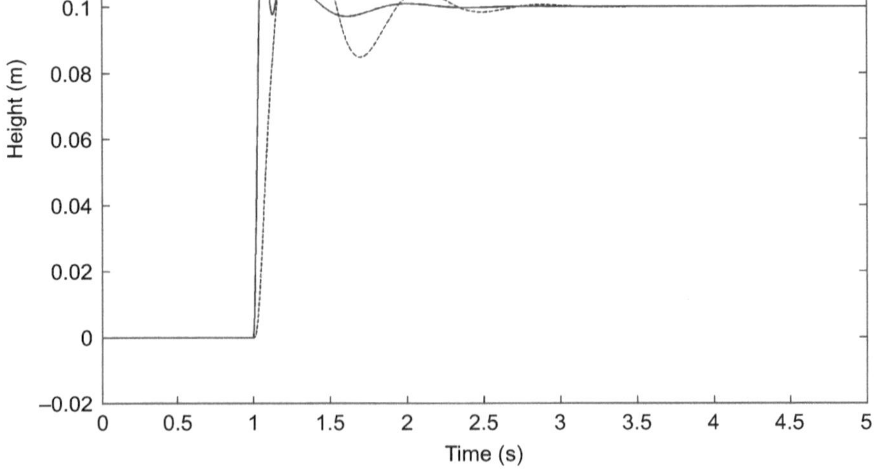

Figure 2.5 The results of the MATLAB simulation for the car suspension system.

2.2.1.1 MATLAB Example: Car Suspension

We now discuss how to represent and simulate the car suspension system in MATLAB. The basic m-file that runs the simulation is shown below, and the simulation results are shown in Figure 2.5. First we show the m-file that runs the basic simulation and describe how the simulation works.

```
% suspension_simulation.m

% Close all figures and clear all variables
close all
clear all

% How long to simulate (in seconds)
t_end = 5;

% Set initial conditions on the system
x1_0 = 0;
x2_0 = 0;
x1_dot_0 = 0;
x2_dot_0 = 0;

% Solve the system equations
[T X] = ode45(@suspension_model,[0 t_end],[x1_dot_0 x1_0 x2_dot_0
x2_0]);

% Save the results in a vector
x1_dot = X(:,1);
x1 = X(:,2);
x2_dot = X(:,3);
x2 = X(:,4);

% Plot the results
plot(T,x1,'k',T,x2,'k--')
xlabel('Time (s)')
ylabel('Height (m)')
title('Wheel Height and Car Height vs. Time')
legend('Wheel Height, x_1','Car Height, x_2')
```

We now discuss the above code in detail and describe each line.

```
close all
clear all
```

These first two lines of the code close figure windows and clear variables so that the simulation is starting with a clean slate.

```
t_end = 5;
```

Next the simulation time is set and stored in the `t_end` variable. This value is used later in the code to solve the system equations for the specified amount of time (5 seconds in this case). Note that MATLAB does not assign units. Rather, it is up to the user to interpret the results and associate appropriate units with the values. We may wish to associate any system of units with (2.3), but in this example and throughout the book, we generally use SI units.

```
x1_0 = 0;
x2_0 = 0;
x1_dot_0 = 0;
x2_dot_0 = 0;
```

In these four lines, the initial conditions are set for the system. In this simulation, the wheel height, car height, and their derivatives set to zero, meaning the system is initially at rest. These values are used to solve the system equations.

```
[T X] = ode45(@suspension_model,[0 t_end],[x1_dot_0 x1_0 x2_dot_0
x2_0]);
```

The next line is the one that solves the system equations given by (2.3). The command `ode45` is MATLAB's differential equation solver that can be used for nonstiff[1] ordinary differential equation problems with a medium order of accuracy. Several other ODE solvers are available (e.g., `ode23` and `ode113`) that may be more appropriate for certain equations.

The `ode45` command uses three parameters to describe how the simulation should run. First, the file containing the system equations (suspension_model.m described below) is included after the @ symbol, indicating a function handle. Second, the simulation time parameters are included in a vector, indicating that the equations should be solved starting at time zero and ending at `t_end`. Finally, the initial conditions are included in a vector as the third parameter in the command.

Last, the `ode45` command returns the results of the solution to the vector `[T X]`. After the solver runs, the time vector is stored in `T`, and the solution

[1] See the series of blog posts by MathWorks' co-founder and chief scientist, Cleve Moler, on the MATLAB ODE solvers (Moler, 2014).

to the differential equations is stored in matrix X. In the case of the car suspension system, T has 437 elements, and X is a 437 × 4 matrix with each row corresponding to the time and each column corresponding to the states $\dot{x}_1$, x_1, $\dot{x}_2$, and x_2 in the order that they appear in the initial conditions and suspension_model.m file.

```
x1_dot = X(:,1);
x1 = X(:,2);
x2_dot = X(:,3);
x2 = X(:,4);
```

These four lines extract the values from the X matrix and put them into separate vectors for each state. This step is not necessary, but it simplifies working with the data.

```
plot(T,x1,'k',T,x2,'k--')
xlabel('Time (s)')
ylabel('Height (m)')
title('Wheel Height and Car Height vs. Time')
legend('Wheel Height, x_1','Car Height, x_2')
```

These last five lines in the file plot x_1 and x_2 versus time for the 5 s of the simulation, add labels, and format the plot. The results of the plot are seen in Figure 2.5. In the plot, notice that the wheel height and car height are at zero until 1 s, when there is a sudden increase and then they settle to a value of 0.1. The fact that the wheel and car heights are both zero does not mean they are at the same height but rather that they are in their equilibrium position with the springs not stretched or compressed beyond what gravity would do.

The jump in values at 1 second is a result of changing the input $r(t)$ from 0 to 0.1. There was nothing in the main simulation program discussed earlier that set the value of r, so where does this occur? The answer lies in the other file specifying the system model, suspension_model.m shown below.

```
function dy = suspension_model(t,y)

% Define the model parameters
m1 = 70;          % kg
m2 = 350;         % kg
k1 = 176000;      % N/m
k2 = 27000;       % N/m
b = 2500;         % Ns/m
```

```
% Define the input r(t)
t_start = 1;
if (t < t_start)
    r = 0;
else
    r = 0.1;
end

dy = zeros(4,1);

dy(1) = -(b/m1)*y(1) - ((k1+k2)/m1)*y(2) + (b/m1)*y(3) + (k2/m1)*
y(4) + (k1/m1)*r;
dy(2) = y(1);
dy(3) = (b/m2)*y(1) + (k2/m2)*y(2) - (b/m2)*y(3) - (k2/m2)*y(4);
dy(4) = y(3);
```

We now describe this m-file line-by-line and how it relates to the system model in (2.3).

```
function dy = suspension_model(t,y)
```

The first line of the file defines it as a function and gives its name (`suspension_model`), which is used in the `ode45` command. The vectors `t` and `y` hold the simulation time parameters `[0 t_end]` and initial conditions `[x1_dot_0 x1_0 x2_dot_0 x2_0]`, respectively.

```
m1 = 70;        % kg
m2 = 350;       % kg
k1 = 176000;    % N/m
k2 = 27000;     % N/m
b = 2500;       % Ns/m
```

These five lines give the values to the constants used in the model.[2]

```
t_start = 1;
if (t < t_start)
    r = 0;
else
    r = 0.1;
end
```

[2] The numbers used in the simulation were obtained from Dixon (2007) and Yong (2008).

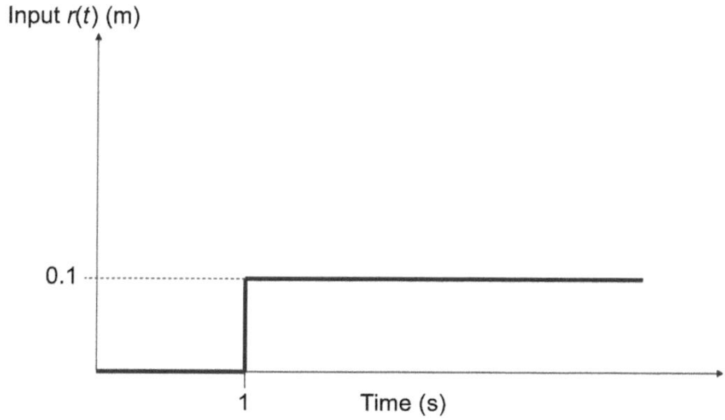

Figure 2.6 The input signal for the simulation of the car suspension.

The `if` statement defines the input $r(t)$. In this case, we wish to have $r(t)$ be a delayed step function with a value of 0.1 after 1 second, as shown in Figure 2.6. The transition time is set in the variable `t_start`, and when the ODE solver time is less than this value, the input is zero. When the time exceeds 1 s, the input is 0.1. These input values are used in the system equations that follow.

```
dy = zeros(4,1);
```

This line initializes the size of `dy` as appropriate. The vector `dy` stores the derivatives of each state variable, which get modified according to (2.3). In this case, the system is fourth order, and the vector is created with four elements. Initially, the values are set to zero, but these will be overwritten with the initial values of the system when the solver runs.

```
dy(1) = -(b/m1)*y(1) - ((k1+k2)/m1)*y(2) + (b/m1)*y(3) + (k2/m1)*
y(4) + (k1/m1)*r;
dy(2) = y(1);
dy(3) = (b/m2)*y(1) + (k2/m2)*y(2) - (b/m2)*y(3) - (k2/m2)*y(4);
dy(4) = y(3);
```

These next four lines specify the system using (2.3), and we will go through a bit of detail to explain where they came from. The system is a fourth-order system and thus needs four equations in the model, but there are only two equations specified in (2.3). Where do the other two equations come from?

The answer lies in the concept **states of the system,** which is described in more detail in Section 2.4, but we will use it here to our advantage in creating the model equations in MATLAB. Two obvious choices for states are x_1 and x_2, the wheel height and car height, respectively. We can generate two more states by taking their derivatives, $\dot{x}_1$ and $\dot{x}_2$. Why not take another derivative to get two more states $\ddot{x}_1$ and $\ddot{x}_2$? One reason is that the system is fourth order, and when we took the first derivative, that gave us four states, and no more are needed. Another reason is that taking more derivatives will cause problems in the procedure we are about to outline.

The model equations in the function file need to provide update equations for each of the system's states with respect to the system states and inputs only. In the case state x_1, we need to tell MATLAB how x_1 changes with time; this is done by specifying its derivative in terms of the states and inputs. In other words,

$$\dot{x}_1 = \dot{x}_1 \tag{2.4}$$

This may seem like a trivial equation, but it needs to be interpreted properly. The left side represents the derivative of x_1, where x_1 is a state. The right side represents the derivative of x_1, which is itself a state of the system.

The update equation for the state $\dot{x}_1$ is not trivial and a bit more complicated. How does $\dot{x}_1$ change with time? We specify it in terms of its derivative,

$$\ddot{x}_1 = -\frac{b}{m_1}\dot{x}_1 - \frac{k_1 + k_2}{m_1}x_1 + \frac{b}{m_1}\dot{x}_2 + \frac{k_2}{m_1}x_2 + \frac{k_1}{m_1}r \tag{2.5}$$

which is obtained by solving the first equation in (2.3) for $\ddot{x}_1$. Similarly, we can go through the same process for x_2. The update equation for x_2 is

$$\dot{x}_2 = \dot{x}_2 \tag{2.6}$$

and the update equation for $\dot{x}_2$ is

$$\ddot{x}_2 = -\frac{b}{m_2}\dot{x}_2 - \frac{k_2}{m_2}x_2 + \frac{b}{m_2}\dot{x}_1 + \frac{k_2}{m_2}x_1 \tag{2.7}$$

We can now see why $\ddot{x}_1$ and $\ddot{x}_2$ cannot be states. The update equations would need to involve $\dddot{x}_1$ and $\dddot{x}_2$, which do not show up anywhere in the derivation of the model.

For ease of notation, let's define

$$\begin{aligned} y_1 &= \dot{x}_1 \\ y_2 &= x_1 \\ y_3 &= \dot{x}_2 \\ y_4 &= x_2 \end{aligned} \tag{2.8}$$

Then the update equations become

$$
\begin{aligned}
\dot{y}_1 &= -\frac{b}{m_1}y_1 - \frac{k_1+k_2}{m_1}y_2 + \frac{b}{m_1}y_3 + \frac{k_2}{m_1}y_4 + \frac{k_1}{m_1}r \\
\dot{y}_2 &= y_1 \\
\dot{y}_3 &= -\frac{b}{m_2}y_3 - \frac{k_2}{m_2}y_4 + \frac{b}{m_2}y_1 + \frac{k_2}{m_2}y_2 \\
y_4 &= y_3
\end{aligned}
\tag{2.9}
$$

These final equations are the ones that appear in the model file. Note that the choice of y_1, y_2, y_3, and y_4 was arbitrary. We could have equivalently set them to any order of $\dot{x}_1$, x_1, $\dot{x}_2$, and x_2. We will see in Section 2.4 that there are conventions about how to define the state variables that make certain forms of model immediately identifiable and easy to study.

The car suspension system is an example of a linear system with one input. The system is described by a set of linear differential equations, meaning that the equations are linear combinations of the states and their derivatives. Furthermore, these differential equations obey the principles of scaling and additivity as discussed in the previous chapter. We contrast this with an example of a nonlinear system and the derivation of its nonlinear differential equations.

Consider a kinematic model for a car. In this model, we assume there is no slip on the wheels, and the vehicle is traveling in the x, y plane with no vertical movement. We also consider only the movement of the vehicle and not acceleration. Under many conditions (e.g., on dry roads and when the car is being driven within its handling limits), these assumptions are valid, and the kinematic model's behavior closely matches the actual car. Given these assumptions, the mathematical model can be derived starting with the coordinate definitions shown in Figure 2.7.

The exact position and orientation of the car in the global coordinate system can be described by four variables. The (x, y) coordinates give the location of the center of the rear axle. The car's angle with respect to the x-axis is given by θ. The steering wheel's angle with respect to the car's longitudinal axis is given by φ. The state of the car can be described by these four variables, and thus it is a fourth-order system requiring four update equations.

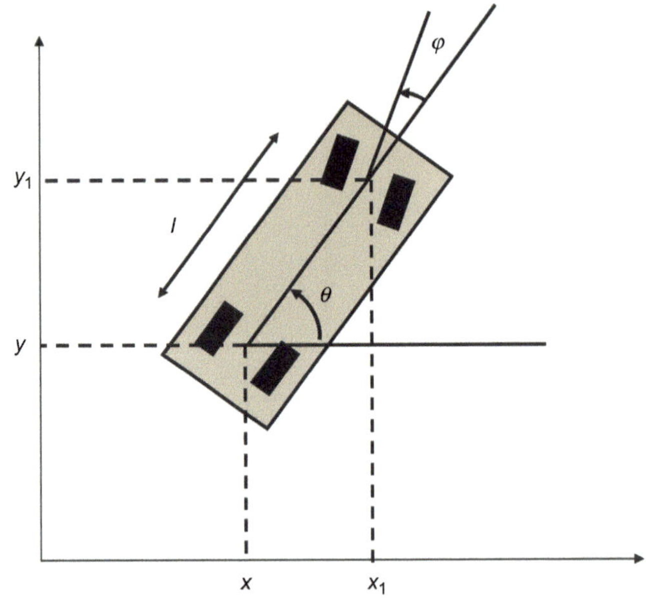

Figure 2.7 The coordinate system for the car model.

From the no-slip constraints, the instantaneous velocity of the car in the x and y directions is given as

$$\begin{aligned}\dot{x} &= v_1 \cos\theta \\ \dot{y} &= v_1 \sin\theta\end{aligned} \tag{2.10}$$

where v_1 is the linear velocity of the rear wheels. The location of the center of the front axle (x_1,y_1) is given by

$$\begin{aligned}x_1 &= x + l\cos\theta \\ y_1 &= y + l\sin\theta\end{aligned} \tag{2.11}$$

and the velocity of this point is given by

$$\begin{aligned}\dot{x}_1 &= \dot{x} - l\,\dot{\theta}\sin\theta \\ \dot{y}_1 &= \dot{y} + l\,\dot{\theta}\cos\theta\end{aligned} \tag{2.12}$$

Now if we apply the no-slip condition to the front wheels, meaning there can be no velocity component perpendicular to the direction of wheel travel, then

$$\dot{y}_1 \cos(\theta + \varphi) = \dot{x}_1 \sin(\theta + \varphi) \tag{2.13}$$

Substituting $\dot{x}_1$ and $\dot{y}_1$ from (2.12) into (2.13) and solving for $\dot{\theta}$ gives

$$\dot{\theta} = \frac{\tan \varphi}{l} v_1 \tag{2.14}$$

Then the complete model for the car is

$$\begin{aligned} \dot{x} &= v_1 \cos \theta \\ \dot{y} &= v_1 \sin \theta \\ \dot{\theta} &= \frac{\tan \varphi}{l} v_1 \\ \dot{\varphi} &= v_2 \end{aligned} \tag{2.15}$$

This is a two-input system in which v_1 is the linear velocity of the rear wheels and v_2 is the angular velocity of the steering wheels.

2.2.1.2 Simulink Example: Kinematic Car Model

Below is a simulation of the kinematic car model using Simulink. This system could be modeled in MATLAB using the above procedure for the suspension system, but instead we will demonstrate the power of Simulink. The Simulink model for the system is shown in Figure 2.8. The model in the figure is annotated to indicate the location of x, y, θ, φ, l, v_1, and v_2. Creating and running the model in Simulink is easier than writing the MATLAB code and gets the same results. However, the MATLAB code offers much more flexibility when system interactions become more complex.

The results of running the Simulink simulation for 10 s are shown in Figures 2.9 and 2.10. The length of the car is set to 1 m, the velocity input v_1 is set to a constant of 5 m/s, and the steering velocity input v_2 is set to a sinusoid of amplitude 0.5 rad/s and frequency 1 rad/s. As with the previous MATLAB example, units are not associated the values but must be interpreted by the user. In Figure 2.9, all the states are plotted versus time using "Scope" blocks. In Figure 2.10, the x,y position of the car is shown having been plotted using the "XY graph" block. The initial position of the car is at (0,0) facing along the x-axis and with straight front wheels because the initial conditions of each integrator was set to zero (the default value). Changing the initial position of the car is obtained by setting the integrator initial conditions.

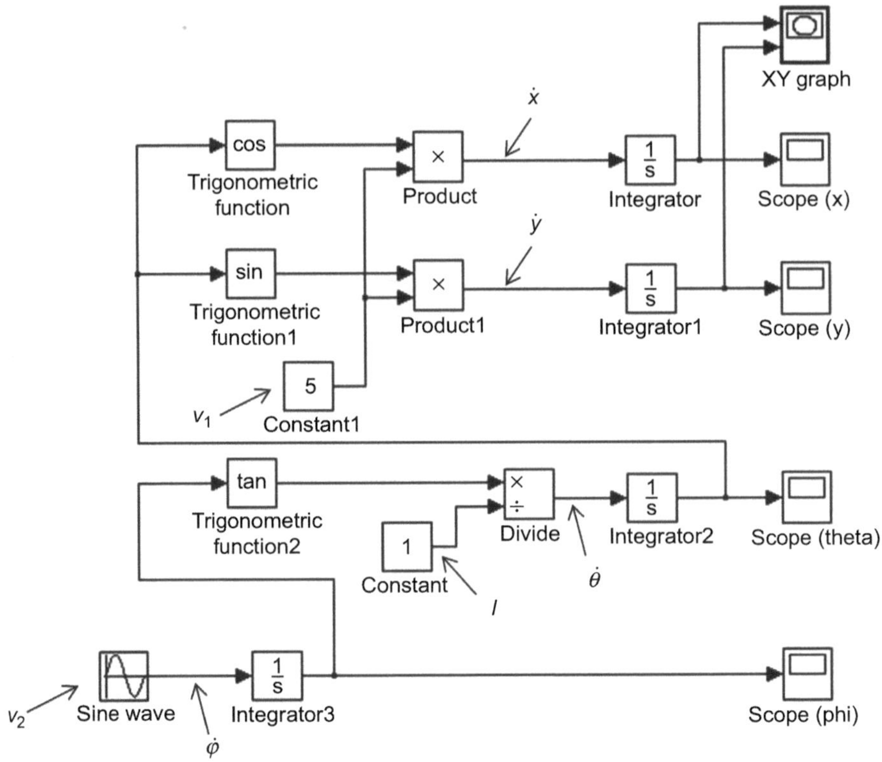

Figure 2.8 The Simulink model for the kinematic car showing the locations of the states, inputs, and parameters.

2.2.2 Difference Equations for Discrete-Time Systems

Just as **differential equations** describe the dynamics of continuous-time systems, **difference equations** describe the dynamics of discrete-time systems. Although the principle is the same, difference equations are often easier to derive and simulate because of their discrete, step-by-step nature.

As a first example, let's again consider the bank account with accruing interest as discussed in Chapter 1. In this example, the formula for the amount of money in the account is

$$A_n = A_0\left(1 + \frac{r}{12}\right)^n \tag{2.16}$$

where r is the annual interest rate, n is the number of months, A_0 is the initial deposit, and A_n is the amount of money in the account in month n. This variable n represents a discrete interval of time. If we think about

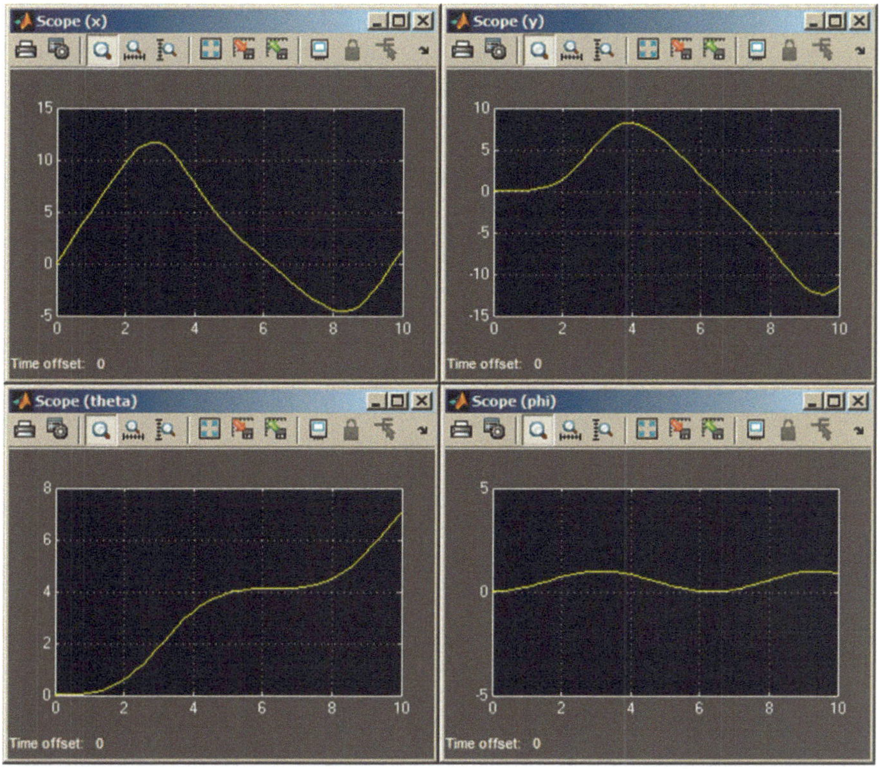

Figure 2.9 x versus time (top left), y versus time (top right), θ versus time (bottom left), and φ versus time (bottom right).

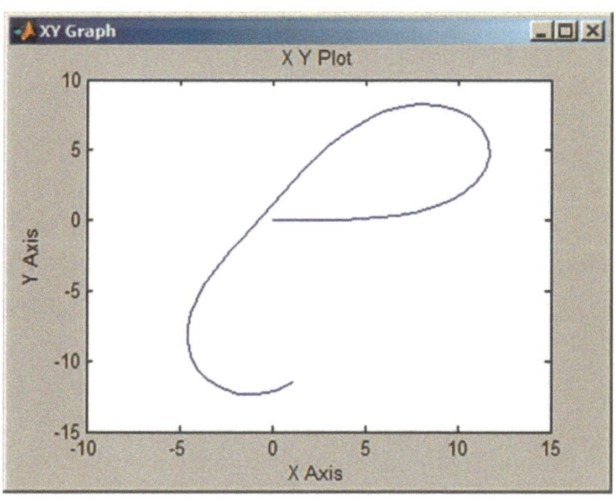

Figure 2.10 The x,y position of the car from 0 to 10 s.

how the amount of money changes at each step, this will give a difference equation to update A, similar to the role differential equations played in the previous two examples. Mathematically, the change is expressed as

$$\begin{aligned} A_n - A_{n-1} &= A_0\left(1+\frac{r}{12}\right)^n - A_0\left(1+\frac{r}{12}\right)^{n-1} \\ &= A_0\left(1+\frac{r}{12}\right)^{n-1}\left(\left(1+\frac{r}{12}\right)-1\right) \\ &= A_{n-1}\left(\frac{r}{12}\right) \end{aligned} \tag{2.17}$$

Solving for A_n yields an update equation.

$$A_n = A_{n-1} + A_{n-1}\left(\frac{r}{12}\right) \tag{2.18}$$

The equation lends itself nicely to a difference equation form, which looks similar but has some notational differences.

$$A[n] = \left(1+\frac{r}{12}\right)A[n-1] \tag{2.19}$$

In this form of the equation, the $[n]$ is used to explicitly show how the expression depends on the discrete-time variable n. It is analogous to showing (t) in the continuous-time case. This equation is already in standard form because the current value of A, denoted by $A[n]$, is isolated and has a coefficient of 1. Note that r is a constant (the interest rate), so the coefficient of $A[n-1]$ is a constant. As with the continuous-time system, this standard form allows us to more efficiently convert to transfer functions and state-space form.

2.2.2.1 MATLAB Example: Bank Account with Interest

We now discuss how to represent and simulate the bank account in MATLAB. You will notice that the discrete-time system simulation file is much simpler than the continuous-time case, mainly because the discrete-time update equation is almost exactly in the form that can be used in MATLAB, and each entry in the vector represents a discrete time step. One does need to be careful about indices for the vectors, however. In the example, the 0 subscript refers to the initial time, but MATLAB does not allow 0 as an index to its arrays. Rather, array indices start at 1, which can cause confusion between the convention used in mathematical representations and its MATLAB implementation. For this reason, two separate

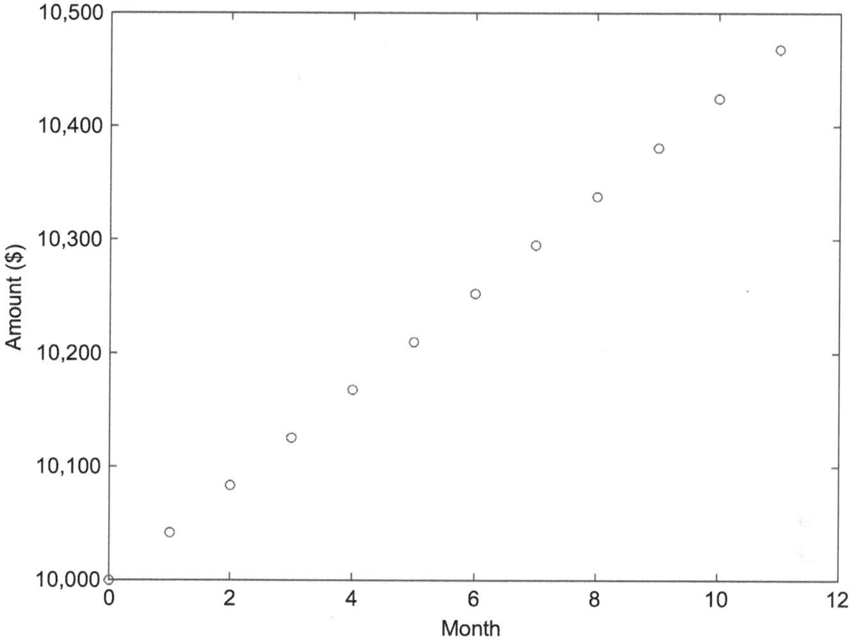

Figure 2.11 Results of the bank account simulation.

array variables are created: n to represent the month as in the equation and k to represent the array index. These two arrays are the same length, but the values in them are offset by 1.

The basic m-file that runs the simulation is shown below, and the simulation results are shown in Figure 2.11.

```
% bank_account.m

close all
clear all

r = 0.05;          % interest rate
n_max = 11;        % how long to run the model
n = 0:n_max;       % define the n vector
A(1) = 10000;      % initial deposit

for k = 2:length(n)
    A(k) = (1+r/12)*A(k-1);
end
```

```
plot(n,A,'o')
y = [10000:100:10500];
set(gca,'YTick',y)
set(gca,'YTickLabel',sprintf('%5.0f|',y))
xlabel('Month')
ylabel('Amount')
```

2.2.3 Models for Hybrid Systems

Now let us consider another example of a real-world discrete-time system: the discretization of a continuous-time system. As discussed in Chapter 1, continuous-time systems are often controlled by some type of computer system (PC based or microprocessor), and these computers are inherently discrete-time systems because they run off a clock and their events occur in discrete time intervals. This is an example of a **hybrid system**, one that has continuous-time and discrete-time parts. The continuous-time part of the system is the mechanical or physical part such as the turning motor or the vibrating mass. The discrete-time part is the computer control that reads sensors and updates command signals at a specific sampling rate.

Hybrid systems are more general than this computer-controlled example. Simply stated, a hybrid system is one that has discrete and continuous dynamics such as the familiar bouncing ball example. If a ball is dropped from some height, it continues to fall until it hits the ground. Up to this point, the dynamics were continuous, and the ball was falling because of gravity. However, when it hits the ground, at that moment its velocity changes direction. In a perfect collision, it would travel with the same speed in the moments before and after impact but with opposite directions. Its new velocity would then be initial conditions for the system equation.

Another example of a hybrid system is a vehicle with a geared transmission. In each of the gears, the vehicle has certain continuous dynamics that relate the fuel input to the speed and acceleration, and these dynamics depend on what gear is selected. Changing gears is a discrete transition between each of these continuous dynamics.

Hybrid systems are an area of much recent research, and readers are encouraged to explore the references at the end of the chapter (Goebel, Sanfelice, & Tell, 2012; van der Schaft & Schumacher, 2000) for more information.

We illustrate the concept of hybrid systems through an example of a computer-controlled car model. To do this, we will revisit the kinematic car model in Simulink and simulate the system as if the velocity commands were coming from a computer at discrete intervals. But first, an overview of the process is provided.

The system simulation follows the steps shown in Figure 2.12. The first step is to set the initial conditions and initial control inputs to the system. With these initial values known, the system equations are solved for T seconds, where T is the sampling time of the system. After this solution is obtained, we see what the endpoint of the solution is and take it to be the initial condition of the system when solving the equations for the following iteration. Also, the system inputs must be chosen (either by closed-loop feedback or an open-loop scheme). With these new initial conditions and inputs, the system equations are solved for another T seconds. This procedure of determining system evolution in T second intervals continues until enough time has elapsed or a stopping condition is met.

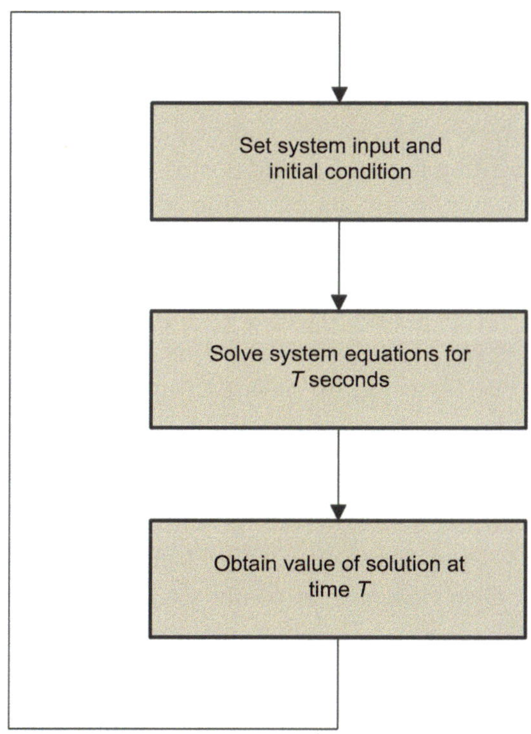

Figure 2.12 Steps in simulating the hybrid vehicle system.

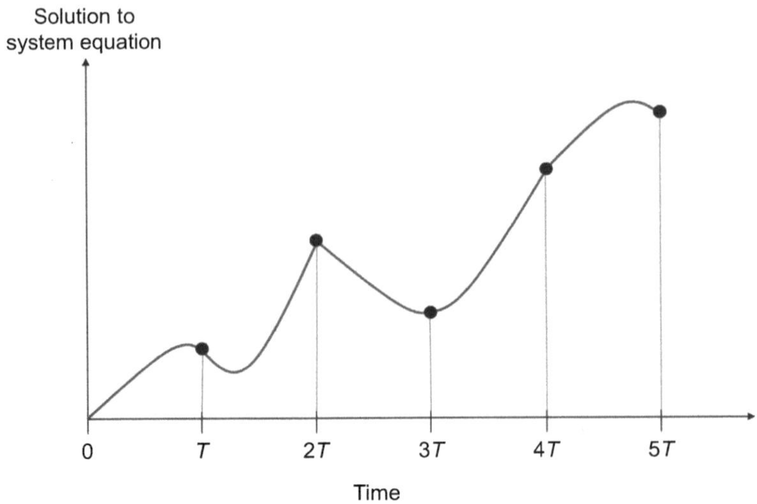

Figure 2.13 Example of a system solution showing the sampling times and solution intervals.

Figure 2.13 shows a possible output of a system simulation. The simulation is carried out over five sample intervals. The curve between each point is the solution to the system equations. Then the endpoint of the curve in each interval is the initial condition to the system equations for the next interval. Because of this condition, the overall solution is continuous, but it may not be differentiable (e.g., at time $2T$ in Figure 2.13). The reason for this lack of smoothness is the inputs to the system. New inputs are applied at each time T, $2T$, $3T$, … , and the inputs themselves are piecewise constant, being held at a constant value during the intervals that the system equations are being solved but may change to new values at each sampling time.

2.2.3.1 MATLAB Example: Computer-Controlled Vehicle Dynamics

We now illustrate the above procedure with an example in MATLAB. This simulation is set up in two parts. The MATLAB code simulates what would run on a microprocessor or computer. It reads the current state of the system from sensors and outputs the commands to the system for control. The Simulink block diagram simulates the system dynamics. The

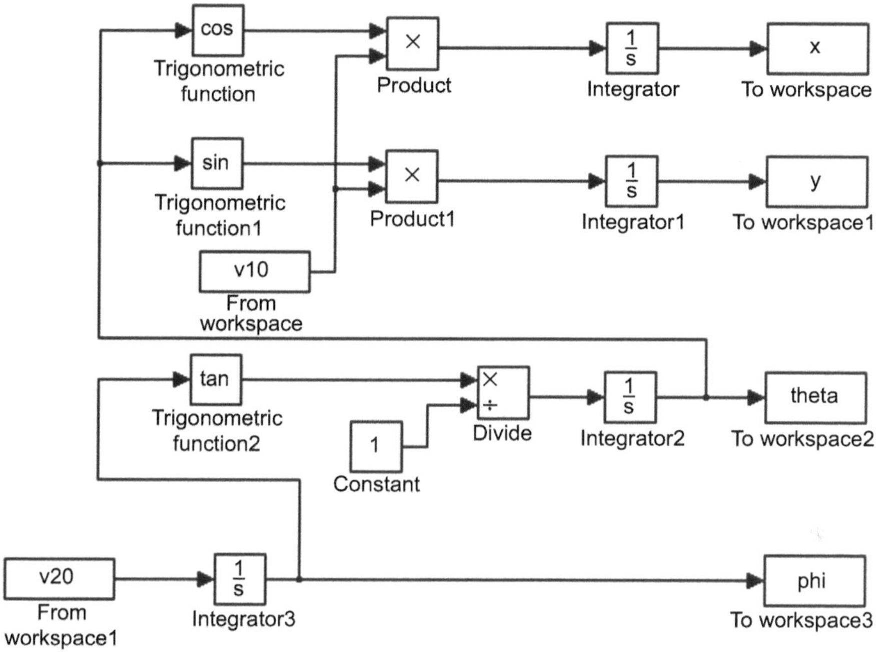

Figure 2.14 The Simulink model for the hybrid car simulation.

MATLAB code calls the Simulink file to initiate the process of solving the system equations.

The MATLAB code is shown below and the Simulink block diagram is shown in Figure 2.14.

```
% hybrid_car.m

close all
clear all

% Initialize constants
T = 0.01;              % Sampling time (seconds)
t_stop = 10;           % Simulation run time (seconds)
t = [0:T:t_stop];      % Time vector
v1 = 5;                % Rear wheel linear speed (m/s)
v2 = 0.5*sin(t);       % Steering wheel speed (rad/s)
```

```
% Set initial conditions
x0 = 0;
y0 = 0;
theta0 = 0;
phi0 = 0;

% Set up state storage arrays
X = zeros(size(t));
Y = zeros(size(t));
THETA = zeros(size(t));
PHI = zeros(size(t));

for k = 1:length(t)

 % Update state storage arrays
 X(k) = x0;
 Y(k) = y0;
 THETA(k) = theta0;
 PHI(k) = phi0;

 % Get current inputs
 v10 = [0 v1];
 v20 = [0 v2(k)];

 % Simulate car movement for time T
 simOut = sim('kinematic_car','StopTime',num2str(T));

 % Store last value of state from simulation
 x0 = simOut.get('x').data(end);
 y0 = simOut.get('y').data(end);
 theta0 = simOut.get('theta').data(end);
 phi0 = simOut.get('phi').data(end);

end

plot(t,X)
title('x vs. t')
xlabel('Time (seconds)')
ylabel('x (meters)')
figure
plot(t,Y)
title('y vs. t')
xlabel('Time (seconds)')
ylabel('y (meters)')
figure
plot(t,THETA)
```

```
title('\theta vs. t')
xlabel('Time (seconds)')
ylabel('\theta (radians)')
figure
plot(t,PHI)
title('\phi vs. t')
xlabel('Time (seconds)')
ylabel('\phi (radians)')
```

For the first part of the code:

```
T = 0.01;              % Sampling time (seconds)
t_stop = 10;           % Simulation run time (seconds)
t = [0:T:t_stop];      % Time vector
v1 = 5;                % Rear wheel linear speed (m/s)
v2 = 0.5*sin(t);       % Steering wheel speed (rad/s)
```

We set the sampling time to 0.01 s, set the end time of the simulation to be 10 s, and define the time vector to go from 0 to 10 s in 0.01-s increments. Also, the inputs are chosen for the entire simulation. The rear wheel speed is set to a constant of 5 m/s, and the steering wheel speed changes according to a slowly varying sinusoid. Using this method, the system inputs are known a priori and thus follow an open loop scheme.

```
% Set initial conditions
x0 = 0;
y0 = 0;
theta0 = 0;
phi0 = 0;
```

Next, the initial conditions are chosen for the system. As in the previous Simulink example, the initial position of the car is at (0,0) facing along the x-axis and with straight front wheels. These variables will be updated at each sampling interval to hold the new initial conditions for the next iteration of the simulation loop.

```
% Set up state storage arrays
X = zeros(size(t));
Y = zeros(size(t));
THETA = zeros(size(t));
PHI = zeros(size(t));
```

These next four lines are for housekeeping purposes. The variables `X`, `Y`, `THETA`, and `PHI` are created to hold the values of the car's four states at each sampling interval so the results can be plotted at the end of the simulation versus time. For this reason, they are made to have the same size as the t vector. The entire array is created before the for loop and set to zero so that MATLAB does not need to resize the array each time through the loop. (This can be a time-saving measure in many situations.)

Now the main for loop begins. First, inside the loop

```
% Update state storage arrays
X(k) = x0;
Y(k) = y0;
THETA(k) = theta0;
PHI(k) = phi0;
```

These four lines update the storage arrays with the current initial condition. (The first time through the loop, these are set to the user's initial condition of all zeros.)

```
% Get current inputs
v10 = [0 v1];
v20 = [0 v2(k)];
```

These following two lines set the inputs for the system equations to be solved in the next line. For the first input, the assignment is redundant because v1 is a constant (5 m/s), and the same value is stored each time through the loop. However, placing the assignment here allows for easy updating of the code for a time-varying v1. The variable v2 is time-varying, and thus the input to the system equation needs to be updated each time through the loop as denoted using the index k, the counter in the for loop.

In both cases, notice the syntax of the assignment statement. The variables v10 and v20 will be passed to the sim command in the next line and will be read into the Simulink model through the "From Workspace" block. This block requires an array containing all the inputs to be applied to the Simulink model and a time index for each one. For this simulation, we are sending a single input value during each iteration; thus, the array has only one row. The zero in the first column is the time index.

```
% Simulate car movement for time T
simOut = sim('kinematic_car','StopTime',num2str(T));
```

The sim command runs the Simulink model contained in the file "kinematic_car.mdl." The command allows the user to specify several parameters in the model, but in this case, we only specify the stop time. The "StopTime" parameter is the one we wish to define, and the following argument in the list, `num2str(T)`, is used because the command expects the stop time value to be passed as a string. The entire list of parameters can be found in the MATLAB documentation.

The Simulink model called in this command is shown in Figure 2.14. This model is almost identical to the one used in Simulink Example 2.2.1.2, but there are a few key differences. First is the way v_1 and v_2 are generated. Here, they are passed from the MATLAB code as described above using the "From Workspace" block rather than created in Simulink using the "Constant" and "Sine Wave" blocks as in the previous example.

Second, the system states x, y, θ, and φ are returned to the MATLAB program using the "To Workspace" blocks. The labels on the blocks must match the variables that are used in MATLAB. In the previous example, these states were simply plotted using the "Scope" blocks. If the plotting was done within Simulink, then only a portion of the entire trajectory would be shown (for that sampling interval) and overwritten the next time the `sim` command is called.

Finally, the initial conditions of each integrator (the initial conditions of the system) are passed from MATLAB. This is accomplished by double clicking on the "Integrator" block and setting the initial condition to be the variable in MATLAB. The parameters for the x integrator block are shown in Figure 2.15. The name used as the "Initial condition" parameter (`x0`) must match the variable name in MATLAB. This is the key to allowing us to update the initial condition of the system for each iteration by saving the last values of x, y, θ, and φ and using them as the new initial conditions in the next iteration.

Returning to the `sim` command in MATLAB, the output stored in the variable `simOut` is an object containing all the values from the Simulink model simulation. To obtain the values from the simulation, the following object oriented assignment statements are used.

```
% Store last value of state from simulation
x0 = simOut.get('x').data(end);
y0 = simOut.get('y').data(end);
theta0 = simOut.get('theta').data(end);
phi0 = simOut.get('phi').data(end);
```

In these lines, the `.get` allows us to access the variables that were sent back from Simulink using the "From Workspace" block. The variable names from those blocks must appear in the parentheses surrounded by

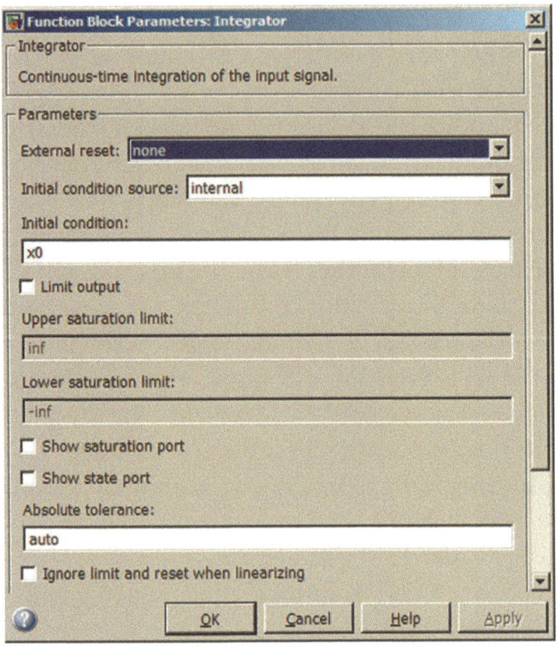

Figure 2.15 The parameters for the integrator block are set in dialog box.

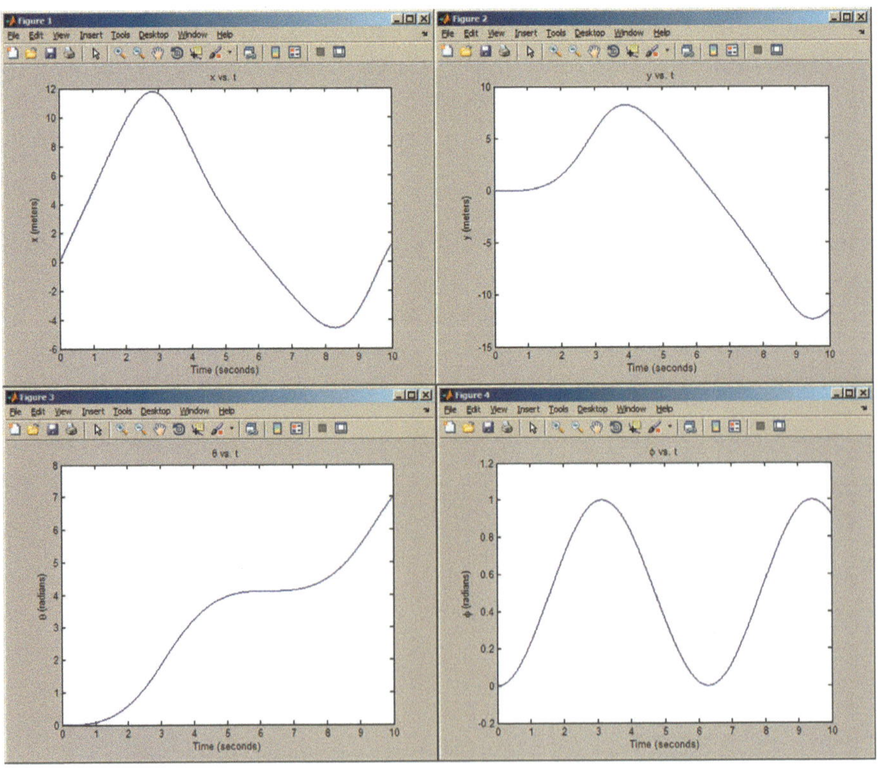

Figure 2.16 The results of the hybrid car simulation with sampling time $T = 0.01$ s.

single quotes. The `.data` returns all the values stored in those variables. In our case, we are only interested in the last value of the solution because this represents the value of the variable at the sampling time and is used as the initial condition for the next iteration. Using `end` as the index of the array allows us to quickly obtain the last value in the array without knowing the exact size of the array.

The final lines of code outside the `for` loop are used to plot the simulation data and format the figures. The results of the simulation are shown in Figure 2.16. Notice that these results match those obtained in the Simulink simulation of the kinematic car shown in Figure 2.9. The reason for the similarity is the hybrid simulation was run with a sampling time `T` of 0.01 s, a short interval that accurately models the physical system. In Figure 2.17, the same results are shown with the sampling time increased to 0.5 s. In this case, the results differ and the piecewise nature of the system can be seen.

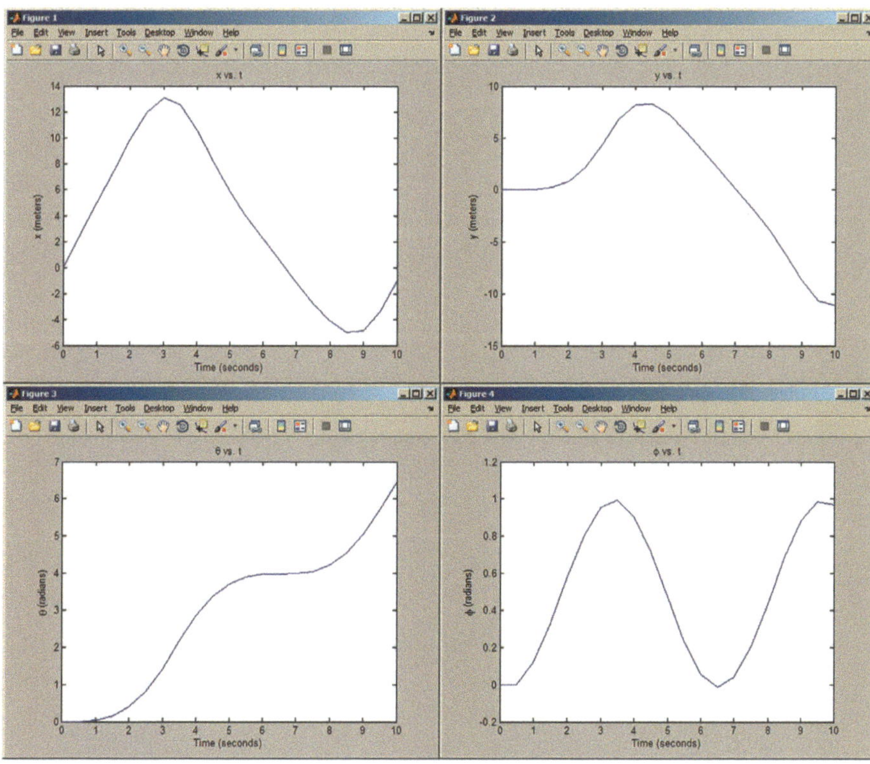

Figure 2.17 The results of the hybrid car simulation with sampling time $T = 0.5$ s.

2.2.4 Flows, Vector Fields, and the Phase Plane

Up to this point, we have looked at the time evolution of the variables in a particular way—plotted with time on the horizontal axis and the variable on the vertical axis. This form of viewing data is intuitive and provides a nice picture of how the individual variables evolve. However, there is another way to look at the data, particularly for second-order systems, which shows relationships between the variables. Known as a **phase plot**, this is a plot commonly showing $\dot{x}$ versus x, where $\dot{x}$ and x are the states of the system. Time is not shown explicitly, but the resulting curves are the system's trajectory parameterized by time. We saw a hint of what a phase plot is in Figure 2.10, which shows the (x,y) position of the car as it evolved over time.

In general, an Nth order continuous dynamical system can be represented by a system of N differential equations.

$$\begin{aligned} \dot{x}_1 &= f_1(x_1, x_2, \cdots, x_N) \\ \dot{x}_2 &= f_2(x_1, x_2, \cdots, x_N) \\ &\vdots \\ \dot{x}_N &= f_N(x_1, x_2, \cdots, x_N) \end{aligned} \tag{2.20}$$

For now, we assume the system inputs are set to zero. We can think of the variables $x_1, x_2, \ldots, x_N$ as **flowing** through their N-dimensional space, starting at some initial state and moving according to their derivatives. This flow is defined by the **vector field** $(f_1, f_2, \ldots, f_N)$ which is used to generate the phase plot as shown in Figure 2.18 for a second-order system. In the figure, one of the states is plotted against the other (x_2 vs. x_1, commonly $x_2 = \dot{x}_1$). The

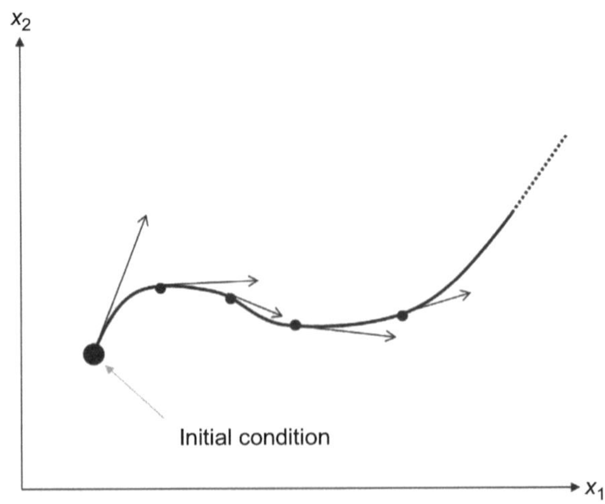

Figure 2.18 A phase plot for a second-order system.

phase plot consists of a smooth curve showing how x_1 and x_2 evolve over time from initial conditions. Each point on the curve has a vector tangent to the curve whose magnitude and direction are defined by $(\dot{x}_1, \dot{x}_2)$, the vector field. If you imagine traveling along the curve, these tangent vectors indicate your velocity. The longer the tangent vector is, the faster you are moving.

Let us illustrate these concepts with a concrete example of a pendulum as shown in Figure 2.19. Assume the pendulum has length l with all its mass m concentrated at the end and that it makes an angle of θ with vertical. We now derive the differential equation for this system.

Assume the only downward force on the pendulum is due to gravity, resulting in a torque pulling the pendulum toward $\theta = 0$. Also assume there is frictional torque acting in the direction opposite the motion and that it is proportional to the pendulum's speed. Summing the torques gives

$$-mgl \sin\theta - b\dot{\theta} = I\ddot{\theta} \tag{2.21}$$

where m is the mass of the pendulum, g is gravitational acceleration, l is the pendulum length, b is the coefficient of friction, and I is the mass moment of inertia. Using $I = ml^2$ and simplifying results in the second-order equation for the pendulum.

$$\ddot{\theta} = -\frac{g}{l}\sin\theta - \frac{b}{ml^2}\dot{\theta} \tag{2.22}$$

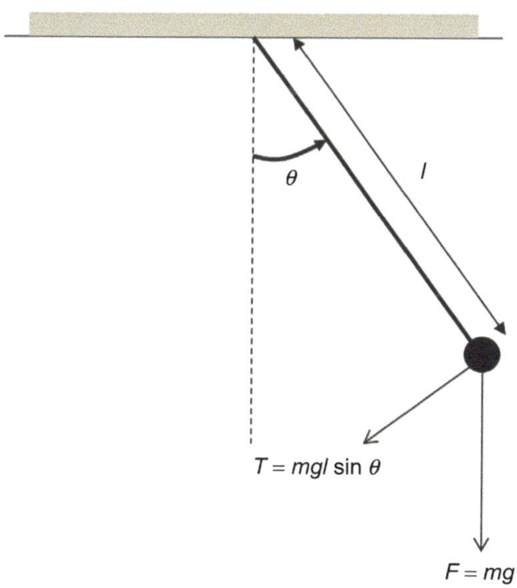

Figure 2.19 The pendulum system.

We next generate a phase plot of $\dot{\theta}$ vs. θ for this system using MATLAB, but first we convert (2.22) into state-space form as we did with the car suspension in MATLAB Example 2.2.1.1.

First define two states

$$\begin{aligned} x_1 &= \theta \\ x_2 &= \dot{\theta} \end{aligned} \tag{2.23}$$

and then obtain the update equations for each state, $\dot{x}_1$ and $\dot{x}_2$

$$\begin{aligned} \dot{x}_1 &= x_2 \\ \dot{x}_2 &= -\frac{g}{l}\sin x_1 - \frac{b}{ml^2}x_2 \end{aligned} \tag{2.24}$$

2.2.4.1 MATLAB Example: Phase Plot of a Pendulum

We now enter the model of the pendulum into MATLAB much like we did for the previous examples. The simulation results are shown in Figures 2.20 and 2.21 for $b = 0$ and Figures 2.22 and 2.23 for $b = 0.1$. The

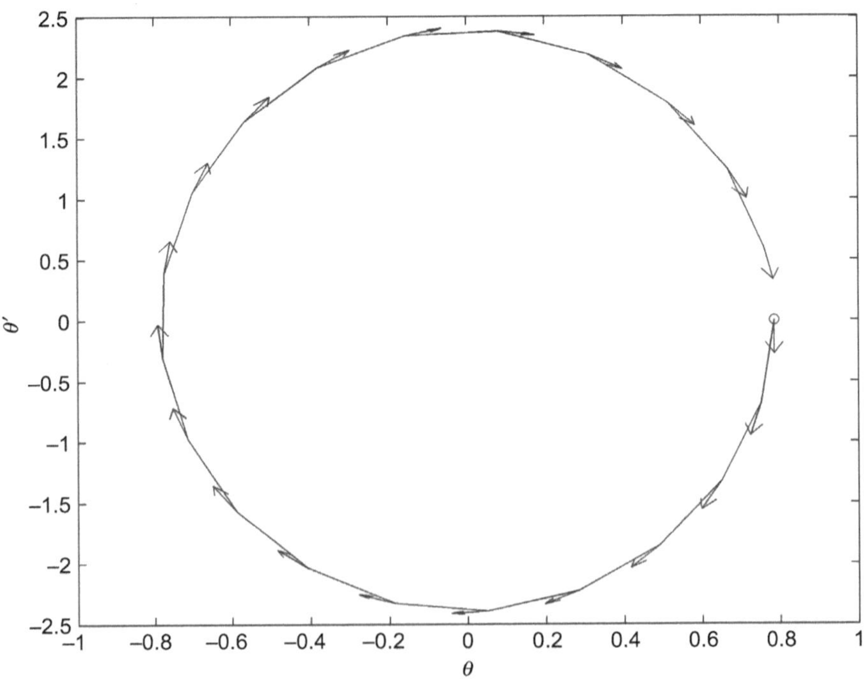

Figure 2.20 Phase plot for the pendulum with $b = 0$.

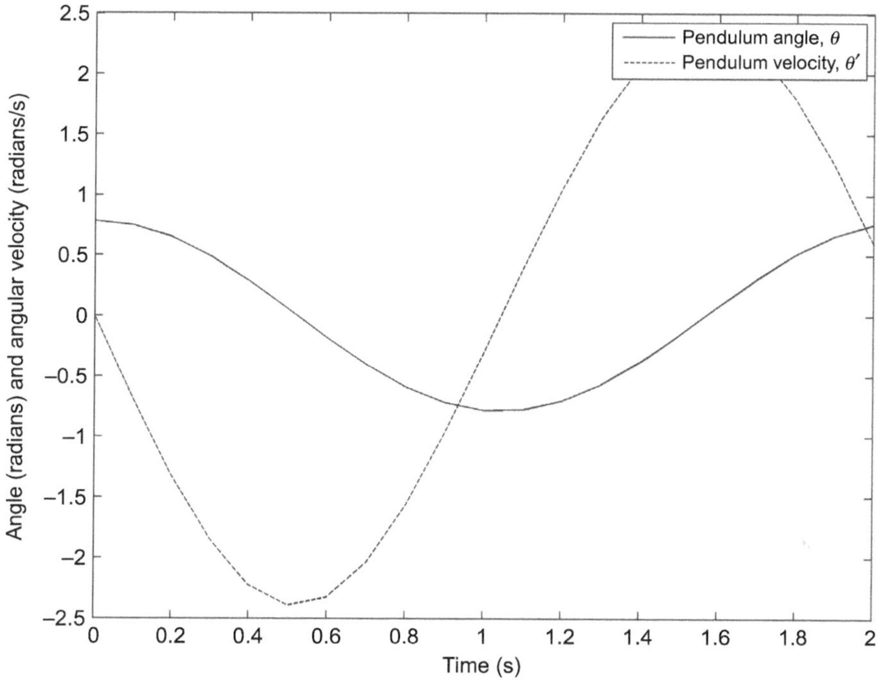

Figure 2.21 Time evolution of pendulum angular position and speed with $b = 0$.

basic m-file that runs the simulation is shown after the figures. This m-file has some differences from the previous examples, and we will discuss those differences.

First let's look at the results in the figures. Figure 2.20 shows the phase plot for the pendulum with no friction ($b = 0$). The initial angle is 45 degrees or 0.79 radians, and the initial speed is zero. The initial condition is denoted by the circle on the plot. At this point, there is a vector pointing down. This vector is defined by $(\dot{\theta}, \ddot{\theta})$, so the component of the vector along the horizontal axis of the plot is 0 because by the initial condition $\dot{\theta} = 0$. The component along the vertical axis is negative, indicating that the pendulum is accelerating back toward $\theta = 0$ due to gravity.

At the point (0,-2.4) on the plot, notice that the pendulum is at its vertical position, and its rotational velocity is at its minimum value (although its absolute value is at a maximum). The vector is pointing to the left, indicating that the velocity is negative and its acceleration is zero. Notice also that all around the curve, the vectors are the same length, meaning that $\sqrt{\dot{\theta}^2 + \ddot{\theta}^2}$ is constant.

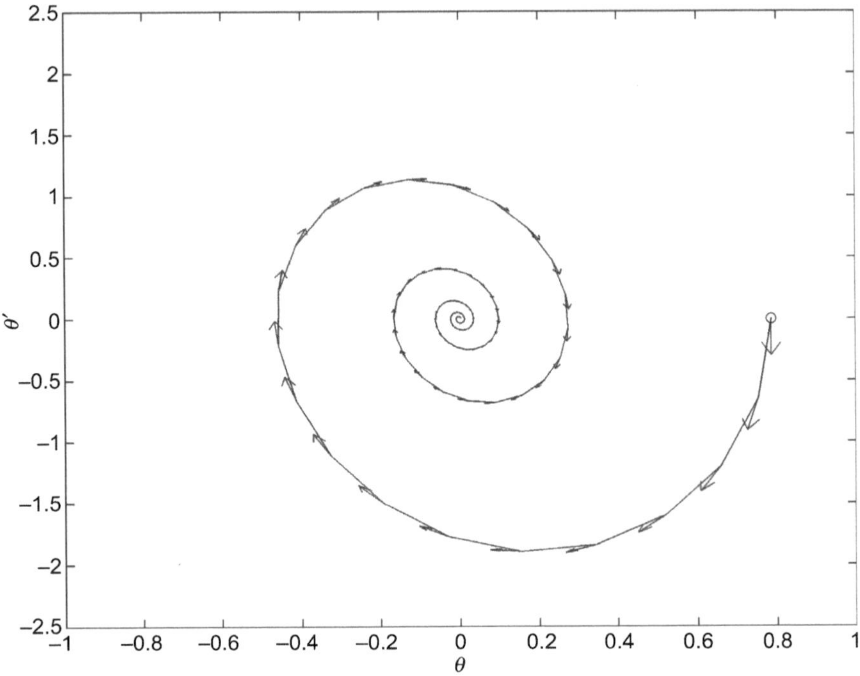

Figure 2.22 Phase plot for the pendulum with $b = 0.1$.

Figure 2.21 shows θ and $\dot{\theta}$ versus time. For each time instant, if the values of θ and $\dot{\theta}$ are plotted as points, the phase plot is generated. Furthermore, the slopes of each of the time plots give the horizontal and vertical components of the vector in the phase plot.

Figures 2.22 and 2.23 show the results when there is friction by setting $b = 0.1$. Starting with the same initial conditions as before, the phase plot (see Figure 2.22) spirals inward toward (0,0). Note that the lengths of the vectors decrease as the curve moves away from the initial point. Eventually, the pendulum stabilizes to $\theta = 0$ and stops moving. The origin is an example of a stable equilibrium point (discussed further in Sections 3.3 and 3.4). In Figure 2.23, the angular position and velocity versus time are seen to have oscillations that eventually dampen to zero.

Now let's discuss how these results were generated. Note that in MATLAB, there are usually several ways to accomplish a desired result, and which one you choose depends on exactly what you want. For example, in this case, the phase plot could have been generated using the `odephas2` as the `OutputFcn` setting in the `odeset` function. However, this option does not produce a desirable

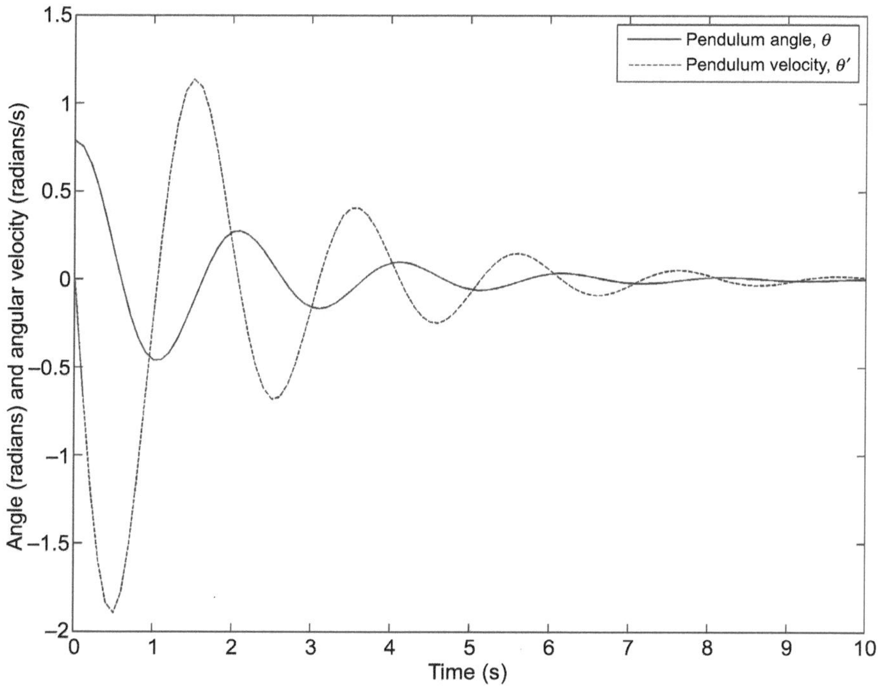

Figure 2.23 Time evolution of pendulum angular position and speed with $b = 0.1$.

looking plot, so the `quiver` function was used instead. This way required more code, but the resulting plots were more appropriate for this discussion.

```
% pendulum_simulation.m

% Close all figures and clear all variables
close all
clear all

% How long to simulate (in seconds)
t_end = 10;
t_sample = 0.1;
t = [0:t_sample:t_end];

% Define the model parameters
m = 0.1;    % kg
l = 1;      % m
b = 0.1;    % kg*m^2/s
g = 9.8;    % m/s^2
```

```
% Set initial conditions on the system
theta_0 = 45*pi/180;
theta_dot_0 = 0;

% Create vectors for storing data
theta = zeros(size(t));
theta_dot = zeros(size(t));

% Solve the system equations
for k = 1:length(t)
theta(k) = theta_0;
theta_dot(k) = theta_dot_0;
[T result] = ode45(@pendulum_model,[0 t_sample],[theta_0
theta_dot_0],odeset,m,l,b,g);
theta_0 = result(end,1);
theta_dot_0 = result(end,2);
end

x = theta;
y = theta_dot;
u = theta_dot;
v = -g*sin(theta)/l-b/(m*l^2)*theta_dot;

% Create the phase plot
quiver(x,y,u,v)
hold on
plot(theta,theta_dot,'b')
plot(theta(1),theta_dot(1),'bo')
axis([-1 1 -2.5 2.5])
xlabel('\theta')
ylabel('\theta\prime')
title('Pendulum Phase Plot')
% Plot the results
figure
plot(t,theta,'k',t,theta_dot,'k-')
xlabel('Time (s)')
ylabel('Angle (radians) and Angular Velocity (radians/s)')
title('Pendulum Angle and Velocity vs. Time')
legend('Pendulum  Angle, \theta','Pendulum  Velocity, \theta^\prime')
```

The above code follows the same basic structure as the hybrid system in the previous example, which simulated the system using sampling. The reason for using the same structure here is that we are "sampling" the trajectory at certain points so that we can plot the vectors at appropriate points along the curve (but not too many to crowd the figure) while still having a smooth curve underneath.

```
% How long to simulate (in seconds)
t_end = 10;
t_sample = 0.1;
t = [0:t_sample:t_end];
```

The code begins with initializing the time vector. The simulation is set to run for 10 s with a sampling time of 0.1 s. As in the hybrid system example, the "continuous" part of the system will be executed for 0.1 s to determine what the values of the system variables are during that time. The ending value of these variables are then used as initial condition for the next sampling period, with these values also being stored in an array for plotting the results later in the code.

```
% Define the model parameters
m = 0.1;      % kg
l = 1;        % m
b = 0.1;      % kg*m^2/s
g = 9.8;      % m/s^2
```

Next, the constants (mass, pendulum length, friction coefficient, and gravitational acceleration) for the problem are defined. This is a change from what was done in previous examples, in which the constants were defined in the model file or Simulink file. The reason for this change is the values must be used to generate the vector field values later in the file. This change also means the `ode45` command must be called in a slightly different way, as we will see below.

```
% Set initial conditions on the system
theta_0 = 45*pi/180;
theta_dot_0 = 0;
```

In these two lines, the initial conditions of the pendulum are set to an angle of 45 degrees and an angular velocity of zero.

```
% Create vectors for storing data
theta = zeros(size(t));
theta_dot = zeros(size(t));
```

The vectors for storing θ and $\dot{\theta}$ are created next. This step is to minimize code run time as before.

```
% Solve the system equations
for k = 1:length(t)
  theta(k) = theta_0;
  theta_dot(k) = theta_dot_0;
  [T result] = ode45(@pendulum_model,[0 t_sample],[theta_0
theta_dot_0],odeset,m,l,b,g);
  theta_0 = result(end,1);
  theta_dot_0 = result(end,2);
end
```

This `for` loop is where the system equations are solved, simulating the "continuous" part of the system as before. There is one difference between this code and the previous examples, and it is how the `ode45` command is called. There are extra entries in the parameter list. The `odeset` function is included to specify that default parameters are used. Then the last set of parameters includes the constants for the system (`m`, `l`, `b`, and `g`). The pendulum_model.m file is shown after the discussion of the main code.

```
x = theta;
y = theta_dot;
u = theta_dot;
v = -g*sin(theta)/l-b/(m*l^2)*theta_dot;
```

In these four lines, the variables needed for the `quiver` command to generate the phase plot are created. The variable names `x`, `y`, `u`, and `v` were chosen to match the documentation for the command. The values in `x` and `y` are plotted to create the trajectory curve. The values in `u` and `v` are the vector fields from (2.24) and are used to plot the arrows along the curve. Because of the way the state variables were defined in (2.23), `y` and `u` happen to be the same.

```
% Create the phase plot
quiver(x,y,u,v)
hold on
plot(theta,theta_dot,'b')
plot(theta(1),theta_dot(1),'bo')
axis([-1 1 -2.5 2.5])
xlabel('\theta')
ylabel('\theta\prime')
title('Pendulum Phase Plot')
```

Next, the quiver command plots the points with their tangent vectors attached as arrows. There are several options for specifying how the plot looks, including a scaling factor to adjust the lengths of the vector arrows. Following the quiver plot, the full trajectory is plotted in blue ("`b`") underneath the arrows, and the initial condition is plotted as "o" in blue as indicated by the "bo" parameter in the plot command. Then the axes are scaled to appropriate values to show the whole plot and labeled.

The model file is shown below.

```
function dx = pendulum_model(t,x,m,l,b,g)

dx = zeros(2,1);

dx(1) = x(2);
dx(2) = (-g/l)*sin(x(1))-(b/(m*l^2))*x(2);
```

This m-file model is similar to the one used in the car suspension example, with one major difference. The constants for the problem are passed to the function as parameters rather than defined in the file so that they can be used in the main m-file.

We have only introduced what phase plots are and how to produce them in MATLAB. However, these plots are rich with information about system behavior. In Chapter 3, we will see how they can be useful tools to study stability and equilibrium points in dynamical systems.

2.3 TRANSFER FUNCTIONS

2.3.1 Overview

So far we have dealt exclusively with systems in the time domain. We have used differential equations and difference equations to mathematically represent how a system behaves, and we have plotted variables versus time and generated phase plots. However, there is another way to mathematically represent systems that is a bit more abstract but holds much information.

A **transfer function** (or system function) is a frequency domain representation of a dynamical system. Before giving going further, let us first express three assumptions that we will use when discussing transfer functions.

1. Transfer functions are used for linear time-invariant systems. Nonlinear or time-varying systems need different analysis techniques.
2. Transfer functions assume the system is initially at rest (zero initial conditions). An example of trying to use transfer functions with nonzero initial conditions (and the associated difficulties) will be given.

3. Transfer functions describe behavior between a single input and a single output. Multi-input and multi-output systems have more than one transfer function to describe the various input—output relationships.

Simply stated, a transfer function of a continuous-time system is defined by

$$H(s) = \frac{Y(s)}{X(s)} \tag{2.25}$$

where $X(s)$ and $Y(s)$ are the Laplace transforms of the system input $x(t)$ and output $y(t)$ respectively.

For a discrete-time systems, the transfer function is defined by

$$H(z) = \frac{Y(z)}{X(z)} \tag{2.26}$$

where $X(z)$ and $Y(z)$ are the z-transforms of the system input $x[n]$ and output $y[n]$, respectively.

We will not discuss Laplace and z-transform theory in detail, nor will we derive many of the relationships and characteristics. In the following sections, we will discuss what the transforms and transfer functions are, how they are used, and apply them to various examples.

2.3.2 Laplace Transforms for Continuous-Time Systems

Laplace transforms are typically introduced in a course on differential equations, and they are used to provide an alternative method for solving differential equations using algebraic methods. This method is quite useful in dynamical system theory in which systems are described by differential equations. If we know the input to the system $x(t)$ and the transfer function $H(s)$, we can determine the output $y(t)$ according to the path shown in Figure 2.24. In the figure, if $x(t)$ is known, we take the Laplace transform, denoted by $\mathscr{L}$. Then we multiply the result by $H(s)$ and take the inverse Laplace transform, denoted by $\mathscr{L}^{-1}$, to get $y(t)$. Although more steps are required than directly solving the differential equation, the steps are simpler, and the many properties of Laplace transforms may be used to make the problem even simpler.

The definition of a Laplace transform of a continuous-time signal $x(t)$ is

$$X(s) = \int_{-\infty}^{\infty} x(t)e^{-st}dt \tag{2.27}$$

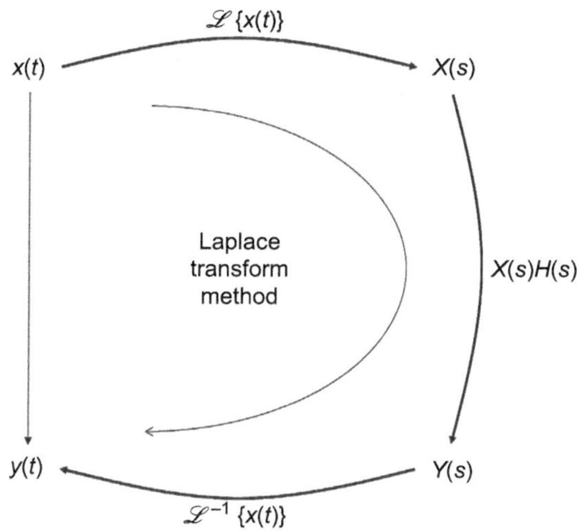

Figure 2.24 Steps in the Laplace transform method of solving differential equations.

In this definition, s is the Laplace variable and is a complex number typically denoted by $s = \sigma + j\omega$, where j is $\sqrt{-1}$.

The inverse Laplace transform is given by

$$x(t) = \frac{1}{2\pi j} \int_{\sigma - j\infty}^{\sigma + j\infty} X(s)e^{st}ds \tag{2.28}$$

In practice, the definitions in (2.27) and (2.28) are rarely used. Instead, tables of transforms and their properties are commonly used to simplify the solution procedure outlined in Figure 2.24.

For our purposes in finding the transfer functions of systems, we introduce notation and properties that will be useful to us. First is to define the notation for signals and their transforms. Table 2.1 shows three variables

Table 2.1 Notation for Laplace Transforms

Time Domain Signal	Laplace Transform
$x(t)$	$\leftrightarrow X(s)$
$y(t)$	$\leftrightarrow Y(s)$
$\theta(t)$	$\leftrightarrow \Theta(s)$

Table 2.2 Important Properties of Laplace Transforms

Property	**Time Domain Signal** $\leftrightarrow$ **Laplace Transform**
Linearity	$\alpha x_1(t) + \beta x_2(t) \leftrightarrow \alpha X_1(s) + \beta X_2(s)$
Differentiation	$\dfrac{dx(t)}{dt} \leftrightarrow sX(s)$
Time Delay	$x(t-\tau) \leftrightarrow e^{-\tau s}\, X(s)$

we have used thus far, $x(t)$, $y(t)$, and $\theta(t)$. These time domain signals are represented by lower case letters and are shown explicitly to be functions of t. Their Laplace transforms are represented by the same letter capitalized and are functions of s.

Three properties of Laplace transforms will be particularly useful to us as we derive transfer functions: linearity and differentiation. These properties are illustrated in Table 2.2.

As an example of deriving a transfer function, we return to the suspension system model Equations (2.3) repeated here.

$$\begin{aligned}\ddot{x}_1(t) + \frac{b}{m_1}\dot{x}_1(t) + \frac{k_1+k_2}{m_1}x_1(t) - \frac{b}{m_1}\dot{x}_2(t) - \frac{k_2}{m_1}x_2(t) &= \frac{k_1}{m_1}r(t) \\ \ddot{x}_2(t) + \frac{b}{m_2}\dot{x}_2(t) + \frac{k_2}{m_2}x_2(t) - \frac{b}{m_2}\dot{x}_1(t) - \frac{k_2}{m_2}x_1(t) &= 0\end{aligned} \tag{2.29}$$

We wish to find the transfer function of the system from road input $R(s)$ to the car height output $X_2(s)$. In other words, the transfer function is $\frac{X_2(s)}{R(s)}$.

The first step is to convert (2.29) by taking the Laplace transform term by term using the properties shown in Tables 2.1 and 2.2. Because the Laplace transform is a linear operation, the constant multipliers remain constant.

$$\begin{aligned}s^2X_1(s) + \frac{b}{m_1}sX_1(s) + \frac{k_1+k_2}{m_1}X_1(s) - \frac{b}{m_1}sX_2(s) - \frac{k_2}{m_1}X_2(s) &= \frac{k_1}{m_1}R(s) \\ \left(s^2 + \frac{b}{m_1}s + \frac{k_1+k_2}{m_1}\right)X_1(s) - \left(\frac{b}{m_1}s + \frac{k_2}{m_1}\right)X_2(s) &= \frac{k_1}{m_1}R(s)\end{aligned} \tag{2.30}$$

$$s^2X_2(s) + \frac{b}{m_2}sX_2(s) + \frac{k_2}{m_2}X_2(s) - \frac{b}{m_2}sX_1(s) - \frac{k_2}{m_2}X_1(s) = 0 \tag{2.31}$$

Solving (2.31) for X_1

$$X_1(s) = \left(\frac{s^2 + \frac{b}{m_2} s + \frac{k_2}{m_2}}{\frac{b}{m_2} s + \frac{k_2}{m_2}} \right) X_2(s) \tag{2.32}$$

and plugging into (2.30) eliminates X_1, yielding an expression with only X_2 as the output and R as the input.

$$\begin{aligned} &\left(s^2 + \frac{b}{m_1} s + \frac{k_1 + k_2}{m_1} \right) \left(\frac{s^2 + \frac{b}{m_2} s + \frac{k_2}{m_2}}{\frac{b}{m_2} s + \frac{k_2}{m_2}} \right) X_2(s) \\ &- \left(\frac{b}{m_1} s + \frac{k_2}{m_1} \right) X_2(s) = \frac{k_1}{m_1} R(s) \end{aligned} \tag{2.33}$$

Simplifying gives the transfer function:

$$\frac{X_2(s)}{R(s)} = \left(\frac{k_1 b}{m_1 m_2} \right) \frac{s + \frac{k_2}{b}}{s^4 + \left(\frac{b}{m_1} + \frac{b}{m_2} \right) s^3 + \left(\frac{k_2}{m_2} + \frac{k_1 + k_2}{m_1} \right) s^2 + \frac{b k_1}{m_1 m_2} s + \frac{k_1 k_2}{m_1 m_2}} \tag{2.34}$$

This transfer has been put into monic form, meaning the coefficients of the highest order of s in the numerator and denominator are 1. This is accomplished by factoring out the $\left(\frac{k_1 b}{m_1 m_2} \right)$, so it is a multiplier in front of the fraction.

2.3.2.1 MATLAB Example: Transfer Function for the Car Suspension

MATLAB is well equipped to handle systems described by transfer functions. Many of the commands we will use here are part of the Control Systems Toolbox.

To represent the car suspension transfer function, we use (2.34) as shown in the code below.

```
% suspension_system_tf.m

close all
clear all

% Define the model parameters
m1 = 70;        % kg
m2 = 350;       % kg
```

```
k1 = 176000;      % N/m
k2 = 27000;       % N/m
b = 2500;         % Ns/m

% Define the transfer function
num = (k1*b/(m1*m2))*[1 k2/b];
den = [1, (b/m1+b/m2), (k2/m2+(k1+k2)/m1), b*k1/(m1*m2), k1*k2/
(m1*m2)];
sys_tf = tf(num,den)

% Plot the step response
step(sys_tf)
```

This code shows the basics of how to represent the system. First the constants are defined using the same values as in MATLAB Example 2.2.1.1.

After the constants are defined, the system is created using the `tf` command. To use this command, we define the numerator and denominator of the transfer function according to the coefficients of the polynomials in s and store them in two different variables, `num` and `den`. These coefficients are simply stored in a row vector and separated by commas (separation by a space would also work). Notice that the `num` variable has a multiplier in front because of the form of (2.34). After the numerator and denominator are defined, these variables are used as parameters in the `tf` command. Because this line does not end with a semicolon, the result is returned to the command window as shown below.

```
sys_tf =

                1.796e04 s + 1.94e05
  ---------------------------------------------------
  s^4 + 42.86 s^3 + 2977 s^2 + 1.796e04 s + 1.94e05

Continuous-time transfer function.
```

Checking the result is always a good idea to make sure the coefficients were entered correctly. One common source of error is to leave out coefficients. For example, $s^2 + 3$ needs to be entered as `[1, 0, 3]`.

The `step` command produces a plot showing the unit step response for the system. The default for this command is to set the system input to a unit step as shown in Figure 2.25. The resulting system output plot is shown in Figure 2.26. The dotted line in the figure indicates the steady-state value for the output.

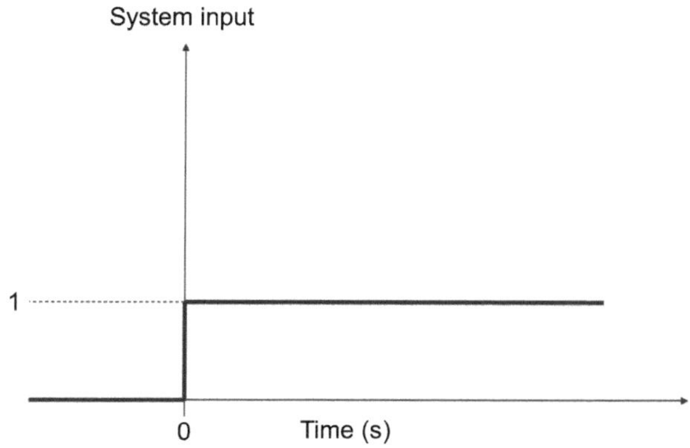

Figure 2.25 The input used in the step command.

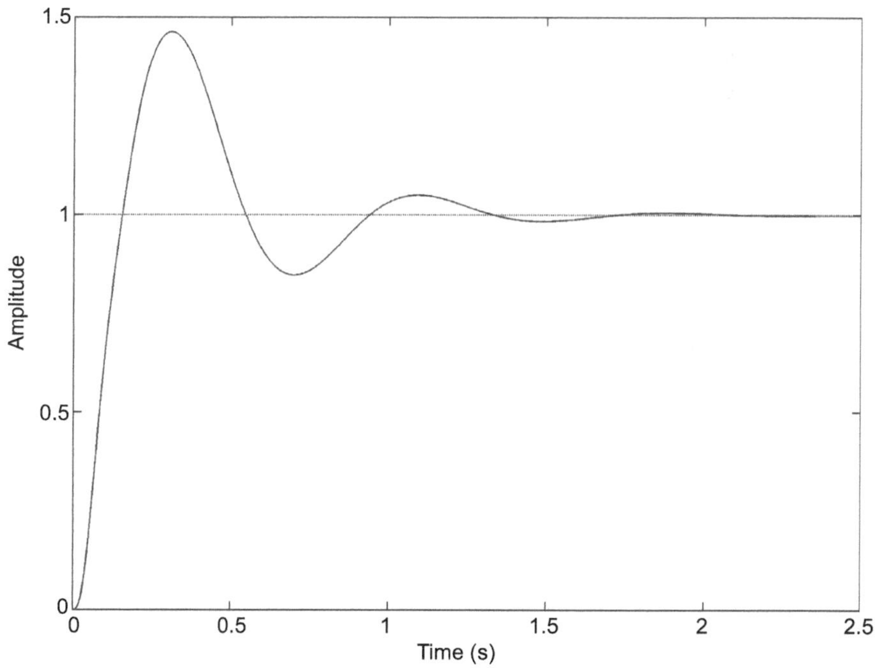

Figure 2.26 The unit step response for the car suspension system.

This step response is almost identical to the one in MATLAB Example 2.2.1.1 except for the scaling and time delay. In the previous example, we set the input $r(t)$ to a step of height 0.1, not 1, and it transitioned from 0 to 0.1 after 1 s. If we want to replicate that input, we can use the `lsim` command to plot the system response to arbitrary inputs. To do this, we define a delayed unit step input and then use it as a parameter in the command as follows.

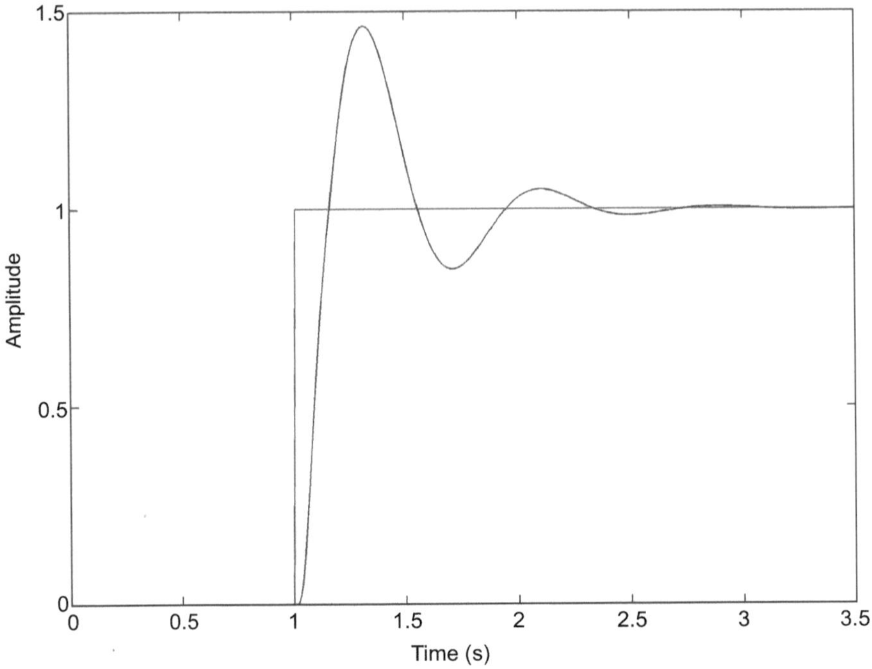

Figure 2.27 The scaled and delayed step response of the car suspension.

```
T = 0.01;
t_end = 3.5;
r_start = 1;
scale_factor = 0.1;
t = [0:T:t_end];
r = [zeros(1,1+r_start/T),scale_factor*ones(1,(t_end-r_start)/T)];
lsim(sys_tf,r,t)
```

In the above code, the input r is defined from 0 to 3.5 s with the transition occurring at 1 s. In the lsim command, the transfer function, input, and time vector must be specified. The resulting plot is shown in Figure 2.27. The input is added to the plot by default as a light gray line.

There is an alternative, and possibly simpler, way to create a transfer function in MATLAB shown below.

```
% Define the transfer function
s = tf('s');
sys_tf = (k1*b/(m1*m2))*(s + k2/b)/(s^4 + (b/m1+b/m2)*s^3 +
      (k2/m2+(k1+k2)/m1)*s^2 + b*k1/(m1*m2)*s + k1*k2/(m1*m2))
```

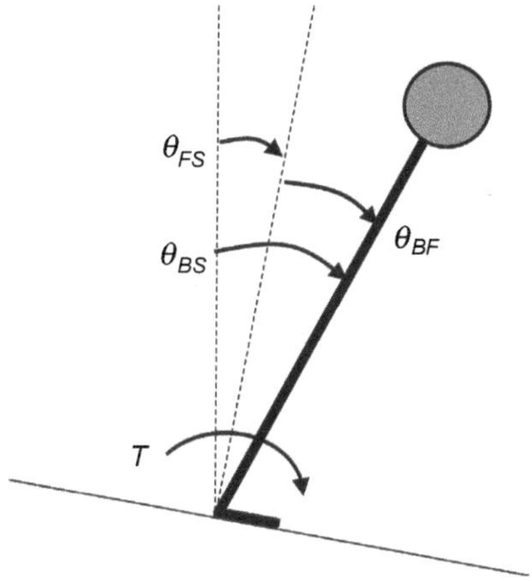

Figure 2.28 Simple model for the human balance system using a single link inverted pendulum.

Using this method, one enters the transfer function directly as shown in (2.34) rather than as coefficients in the numerator and denominator. It is possibly simpler because it is just like the transfer function expression, but it can be much longer, and the chances are greater of making a mistake while typing it into the code. Both methods produce the same result and it's the user's discretion to choose the more suitable one.

As a second example, we give a simple model of the human balance system as a single link inverted pendulum.[3] This system incorporates feedback and we will show how MATLAB can help with the transfer function derivation.

The single link inverted pendulum is shown in Figure 2.28. In the figure, the person is represented by a single pendulum rotating about the ankles. There are three angles associated with the system: θ_{BS}, the body's angle in space; θ_{FS}, the foot's angle in space; and θ_{BF}, the foot's angle with respect to the body. Additionally, the person may apply torque T about the ankle to control balance. For this simple model, we assume the mass m is concentrated at the center of mass located distance l above the feet.

The differential equation describing the motion of the system is

$$mgl \sin \theta_{BS} + T = J\ddot{\theta}_{BS} \tag{2.35}$$

[3] The model is described in Peterka (2003).

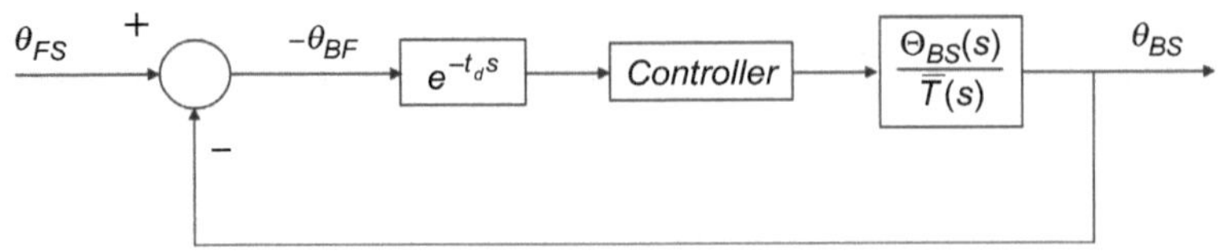

Figure 2.29 A simple model of human balance using proprioception in a feedback system.

Because this is nonlinear system, it does not have a transfer function representation. However, using the linear approximation $\sin\theta_{BS} \approx \theta_{BS}$, the transfer function from T to θ_{BS} is found to be[4]

$$\frac{\Theta_{BS}(s)}{\overline{T}(s)} = \frac{\frac{1}{J}}{s^2 + \frac{mgl}{J}} \tag{2.36}$$

This transfer function describes how torque applied to the ankle affects the body angle. To complete the model, we have to account for how a person might react to changing foot angle θ_{FS}. To accomplish this effect, the system is modeled using feedback as shown in Figure 2.29. This model assumes only proprioception is used for feedback (i.e., joint angles are used to determine body position with respect to itself) and ignores the vestibular and vision systems. The input to the system is the foot's angle in space, which represents the slope on which the person is standing. The output is the body's angle in space, which is controlled by applying torque to the ankle. The difference between these two angles is θ_{BF}, which is sensed using proprioception.

Researchers use this model to help understand the human balance system. One of the key features that they had to identify was the form of the controller. They have found that a proportional-integral (PI) controller works when fitting the model to the data. A PI controller has a transfer function $C(s)$ given by

$$C(s) = k_P + \frac{k_I}{s} \tag{2.37}$$

where k_P and k_I are called the proportional gain and integral gain, respectively. These are constants that must be chosen to match the model to the data.

Also note the left block in the diagram whose transfer function is e^{-st_d}. This block represents a delay time of t_d in the system caused by the person's reaction

4 Because the time domain representation of torque T was already capitalized, the bar notation is used for its Laplace transform.

time. As discussed earlier, a signal delayed by t_d in the time domain corresponds to multiplication by e^{-st_d} when using Laplace transforms. Researchers must also determine t_d to make the model accurately fit their collected data.

Using properties of feedback loops, the overall transfer function of the system from θ_{FS} to θ_{BS} is found to be

$$\frac{\Theta_{BS}}{\Theta_{FS}} = \frac{e^{-st_d}\left(k_P + \frac{k_I}{s}\right)\left(\frac{\frac{1}{J}}{s^2 + \frac{mgl}{J}}\right)}{1 + e^{-st_d}\left(k_P + \frac{k_I}{s}\right)\left(\frac{\frac{1}{J}}{s^2 + \frac{mgl}{J}}\right)} \tag{2.38}$$

2.3.2.2 MATLAB Example: Transfer Function of the Human Balance System

We can use MATLAB to represent and simulate the human balance system shown in Figure 2.29. In this example, we use a new tool to help with some of the work.

So far we have seen two ways to enter transfer functions. One way is to enter the transfer function as a numerator and denominator. In this case, it would mean simplifying (2.38) to a simple fraction with both the numerator and denominator expressed as polynomials of s and creating two vectors in MATLAB to hold the polynomial coefficients. However, these coefficients cannot be found because of the delay component e^{-st_d}. The other way is to define `s` using the `tf` command and then enter the actual polynomials and let MATLAB do the fraction simplification. In this case, we need to do less work than the first method because we are required to know that the feedback system is of the form in (2.38) but do not need to do calculations.

There is a third way to enter the transfer function, and this method puts most of the work onto MATLAB. The `feedback` command lets us enter the individual blocks, and then it does the simplification. The format of the command is

```
feedback(M1,M2)
```

where M1 and M2 show up in the system as shown in Figure 2.30. If not specified, it is assumed that there is negative feedback.

One other item to discuss about the model is the delay component e^{-st_d}. Handling this block is easy when we first define `s` using `tf`. We simply use the MATLAB command `exp` to enter the mathematical expression.

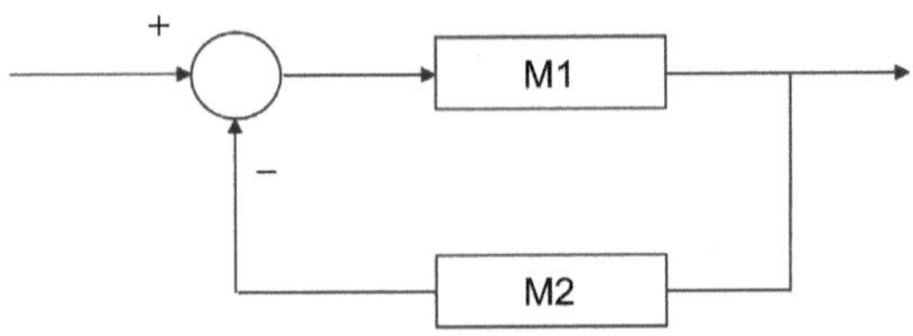

Figure 2.30 The form of the system assumed when using MATLAB's feedback command.

The code to define and simulate the system is shown below.

```
% human_balance.m

close all
clear all

% Define the system constants
td = 0.01;   % s
J = 81.3;    % kg m^2
m = 82;      % kg
g = 9.8;     % m/s^2
L = 0.996;   % m
kp = 0.5;
ki = 0.001;

% Define the system blocks
s = tf('s');
D = exp(-s*td);
C = kp + ki/s;
G = (1/J)/(s^2+m*g*L/J);

sys = feedback(D*C*G,1)
```

In the code, first the constants of the system are defined. Some typical values are used here[5] Next the individual blocks are defined using `s` as the variable, which allows us to type in the mathematical expressions rather than just polynomial coefficients. Finally, the system is defined using the `feedback` command with the series connections of the three

[5] From Goodworth, Mellodge, and Peterka (2014).

blocks D, C, and G in place of M1. The result of this command is shown below.

```
sys =

  a =
                 x1          x2          x3
   x1             0      -2.463  -3.075e-06
   x2             4           0           0
   x3             0           1           0
  b =
                 u1
   x1       0.03125
   x2             0
   x3             0

  c =
                 x1          x2          x3
   y1             0      0.0492    9.84e-05

  d =
                 u1
   y1             0

  (values computed with all internal delays set to zero)

  Internal delays (seconds): 0.01

Continuous-time state space model.
```

Notice that MATLAB gives us something that looks different from what we have seen in previous examples. Rather than return a transfer function, it represents the system with four matrices a, b, c, d. This is a form of state-space representation, and we will discuss this topic in Section 2.4.

2.3.2.3 Systems with Nonzero Initial Conditions

Consider the simple circuit shown in Figure 2.31. Assume the capacitor is initially charged to some nonzero voltage $v_c(0)$.

Applying Ohm's law to the resistor gives

$$i(t) = \frac{v_{in}(t) - v_c(t)}{R} \tag{2.39}$$

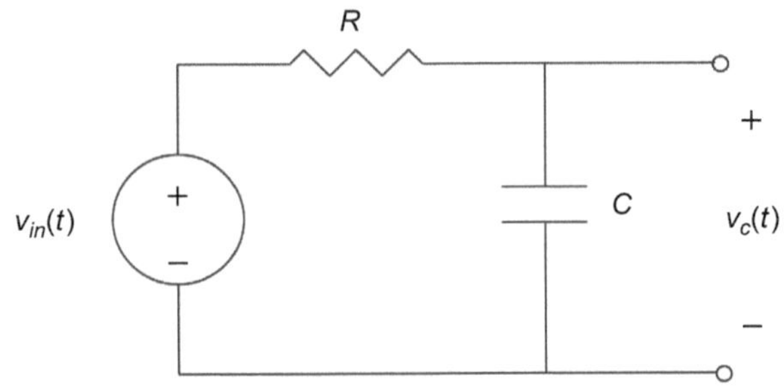

Figure 2.31 A low-pass resistor capacitor (RC) filter circuit.

and taking the Laplace transform of (2.39) results in

$$I(s) = \frac{V_{in}(s) - V_c(s)}{R} \tag{2.40}$$

Furthermore, the relationship between current and voltage in a capacitor is

$$i(t) = C\frac{dv_c(t)}{dt} \tag{2.41}$$

Taking the Laplace transform of (2.41) results in the initial condition on the capacitor appearing

$$I(s) = C(sV_c(s) - v_c(0)) \tag{2.42}$$

Trying to obtain the transfer function $\frac{V_c(s)}{V_{in}(s)}$ is hopeless. Substituting (2.42) into (2.40) and simplifying yields

$$V_c(s) = \frac{\frac{1}{RC}}{s + \frac{1}{RC}} V_{in}(s) - \frac{1}{s + \frac{1}{RC}} v_c(0) \tag{2.43}$$

and shows that the output V_c is a function of both the input V_{in} and the initial condition $v_c(0)$. However, if we assume zero initial conditions, the transfer function becomes

$$\frac{V_c(s)}{V_{in}(s)} = \frac{\frac{1}{RC}}{s + \frac{1}{RC}} \tag{2.44}$$

What if you want to plot the time response of the RC circuit in MATLAB for nonzero initial conditions? Fortunately, there is a simple means to accomplish this using the system's state-space realization. In

Section 2.4, we will show how to carry this out in MATLAB and give a detailed discussion of state-space models.

2.3.3 *z*-Transforms for Discrete-Time Systems

The concepts of *z*-transforms are generally less exposed to mechanical engineers than they are to electrical engineers, who get a heavy dose of them in required undergraduate courses in signals and systems theory, which are then followed by courses in signal processing, communications, and control systems. In particular, digital signal processing (DSP), digital communication systems, and digital control systems rely heavily on *z*-transforms. Mechanical engineers are less likely to study *z*-transforms in their undergraduate coursework, but may encounter them in a course on digital control systems or mathematically oriented courses at the graduate level.

The *z*-transform is to discrete-time systems and difference equations what Laplace transforms are to continuous-time systems and differential equations. Many of the same properties apply that we saw in discussing Laplace transforms. First we will start with the definition of the *z*-transform of a discrete-time signal.

The definition of a *z*-transform of a discrete-time signal $x[n]$ is

$$X(z) = \sum_{n=-\infty}^{\infty} x[n]z^{-n} \tag{2.45}$$

where z is a complex number.

The inverse *z*-transform is given by

$$x[n] = \frac{1}{2\pi j} \oint_C X(z)z^{n-1}dz \tag{2.46}$$

where C is a counterclockwise path that encloses the origin and the region of convergence.

Region of convergence is defined as the set of points in the complex z plane for which the *z*-transform in (2.45) converges. Notice that (2.45) is an infinite sum, so it will only give a finite value under certain conditions for a given discrete-time signal $x[z]$. For the *z*-transform to be complete, the region of convergence must be specified along with $X(z)$.

In practice, as with Laplace transforms, it is more common to use tables than the definitions in (2.45) and (2.46). However, the definition of the *z*-transform is a bit easier to use than the Laplace definition because one can consider multiplication by z^{-n} to be a delay operator, and difference equations used to describe a discrete-time system can often be expressed as a linear combination of delays.

Let us look at an example of a signal and its z-transform to demonstrate these ideas. Suppose $x[n]$ is given by

$$x[n] = \begin{cases} e^{-n}, & n \geq 0 \\ 0, & n < 0 \end{cases} \tag{2.47}$$

Using (2.45), the z-transform is

$$X(z) = \sum_{n=0}^{\infty} e^{-n} z^{-n} \tag{2.48}$$

Note that the lower limit of the summation is zero because $x[n] = 0$ when n is less than 0. Simplifying the expression gives

$$\begin{aligned} X(z) &= \sum_{n=0}^{\infty} (ez)^{-n} \\ &= \sum_{n=0}^{\infty} \left((ez)^{-1}\right)^{n} \\ &= \frac{1}{1-(ez)^{-1}} \end{aligned} \tag{2.49}$$

The last step was obtained using the relationship $\sum_{n=0}^{\infty} x^n = \frac{1}{1-x}$. This relationship only holds if $|x|$ is less than 1; otherwise, the series does not converge. This restriction leads us to the region of convergence (ROC). In the above example, the ROC is

$$\left|(ez)^{-1}\right| < 1 \tag{2.50}$$

or

$$|z| > e^{-1} \tag{2.51}$$

Because z is a complex number, this last inequality says that z must lie outside a circle of radius e^{-1} for the series to converge. Graphically, the region of convergence is the area shown in Figure 2.32.

Thus the complete z-transform is

$$X(z) = \frac{1}{1-(ez)^{-1}}, \qquad |z| > e^{-1} \tag{2.52}$$

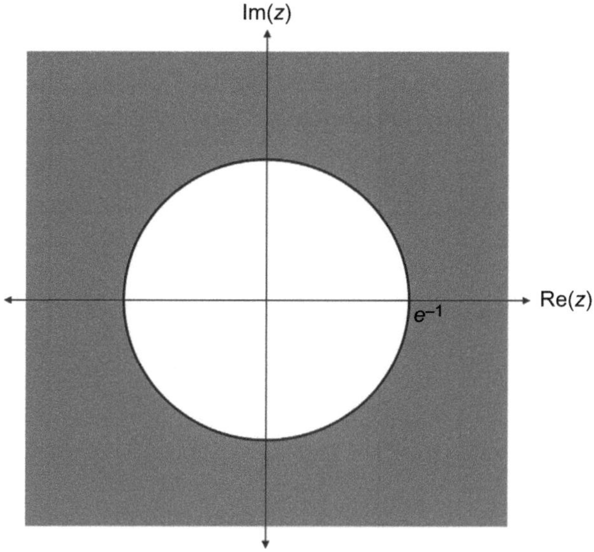

Figure 2.32 The region of convergence for z shown on the complex plane. The z-transform converges for values of z in the shaded region.

As with Laplace transforms, we introduce notation and properties that will be useful to us. First is to define the notation for signals and their transforms. These time domain signals are represented by lower case letters and are shown explicitly to be functions of n. Their z-transforms are represented by the same letter capitalized and are functions of z.

Table 2.3 Notation for z-Transforms

Time Domain Signal	z-Transform
$x[n] \leftrightarrow X(z)$	

Table 2.4 Important Properties of z-Transforms

Property	Time Domain Signal	z-Transform
Linearity	$\alpha x_1[n] + \beta x_2[n] \leftrightarrow \alpha X_1(z) + \beta X_2(z)$	
Time shift	$x[n - n_0] \leftrightarrow z^{-n_0} X(z)$	

Some properties of z-transforms will be particularly useful to us as we derive transfer functions: linearity and time shift. The notation is illustrated in Table 2.3. These properties are illustrated in Table 2.4.

As a simple first example, consider the difference equation in (2.53), and let's work through the process of generating its transfer function.

$$x[n] + 3x[n-1] + 4x[n-2] = y[n] - y[n-2] \tag{2.53}$$

becomes

$$X(z) + 3z^{-1}X(z) + 4z^{-2}X(z) = Y(z) - z^{-2}Y(z) \tag{2.54}$$

and the transfer function $\frac{Y(z)}{X(z)}$ is

$$\frac{Y(z)}{X(z)} = \frac{1 + 3z^{-1} + 4z^{-2}}{1 - z^{-2}} \tag{2.55}$$

or equivalently

$$\frac{Y(z)}{X(z)} = \frac{z^2 + 3z + 4}{z^2 - 1} \tag{2.56}$$

These two forms of transfer function are used in practice, depending on the field of study. In DSP, it is common to express transfer functions for discrete-time systems in terms of negative powers of z as in (2.55) because of their association with delay operations. For control systems, positive powers of z are typically used. As we will see in the MATLAB example later in this section, either convention may be used as long as the user specifies which he or she is following.

The most common discrete-time system that we will see in studying mechanical systems comes about through sampling a process for computer control. As an example, we will work through a thermostat system in detail and see the effects of this sampling on the model.

Consider a computer-controlled heating system represented by the block diagram in Figure 2.33. In this situation, the computer samples the

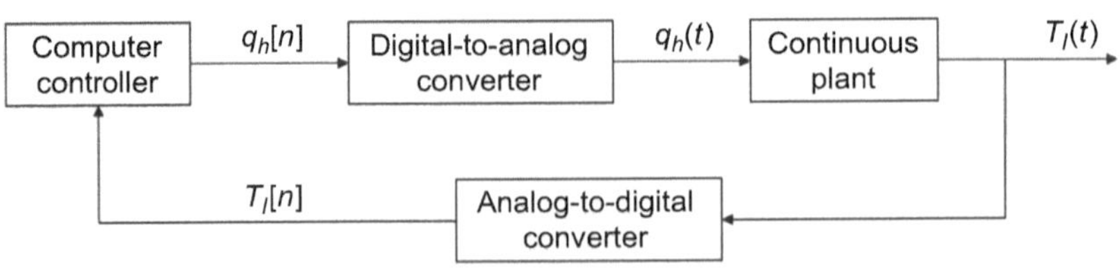

Figure 2.33 A closed loop temperature control system for a room.

room temperature at discrete intervals using an analog-to-digital converter (ADC). Note that the room temperature is denoted by $T_I(t)$ because it is a continuous-time signal. The sampled version of the room temperature coming out of the ADC is $T_I[n]$, a discrete-time signal. Similarly, the computer outputs a discrete-time signal $q_h[n]$, which goes through a digital-to-analog converter (DAC) to produce a continuous-time signal $q_h(t)$.

For our modeling purposes, we assume the ADC has infinite resolution. It simply samples the continuous-time signal and outputs the exact value of the signal at integer multiples of the sampling time t_s. The value of the input signal is captured at each sampling instant so that the relationship between the two signals is

$$T_I[n] = T_I(nt_s) \tag{2.57}$$

This relationship between the signals is shown graphically in Figure 2.34. Figure 2.35 shows how this relationship can be modeled in a block diagram as a switch that closes momentarily every t_s seconds. Note that this model is an idealized version of how the ADC behaves, and it ignores quantization error. In an actual ADC, the sampled value is

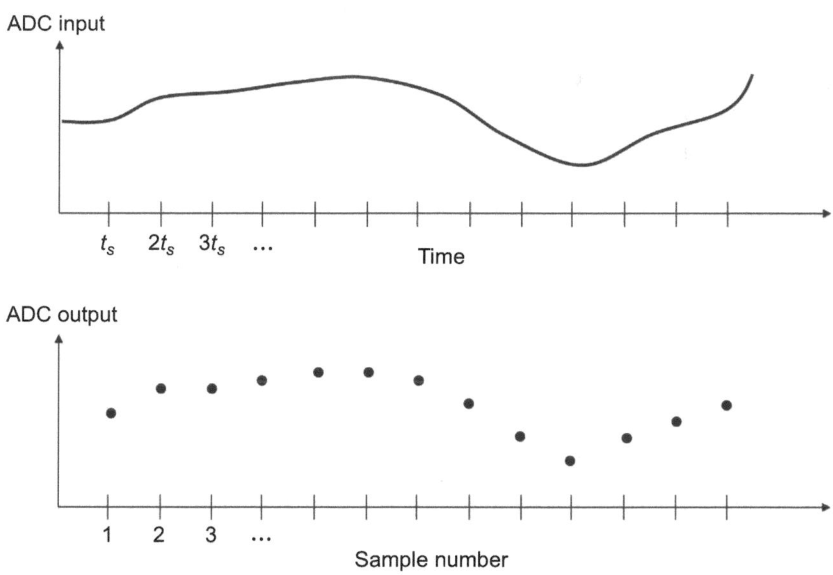

Figure 2.34 The input and output signal for an analog-to-digital converter (ADC).

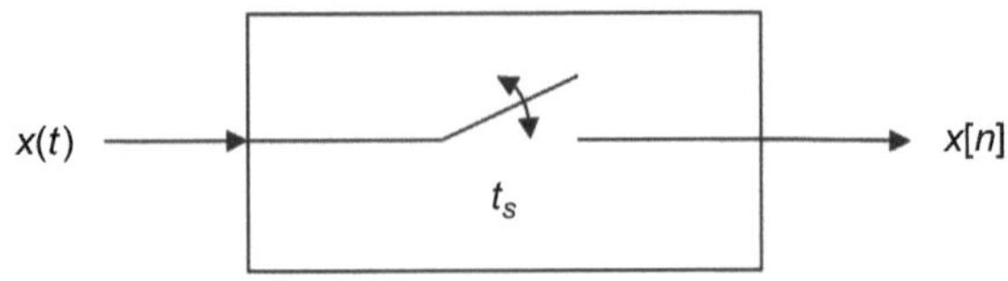

Figure 2.35 The model of an ideal analog-to-digital converter.

represented by a finite number of bits. For example, in 12-bit sampling, the value of the signal is quantized into one of $2^{12} = 4096$ values.

The DAC is modeled differently from an ADC. Typically a DAC is represented using what is called a zero-order hold (ZOH) equivalent model. In this case, the input to a DAC is a discrete-time signal, and the output is the input value held constant for the time between samples. The result of the conversion is a piecewise constant signal as shown in Figure 2.36.

If we consider the input and output to be continuous-time signals, then we can derive the transfer function of the DAC in terms of the Laplace transforms of the input and output signals. For example, if the input is a unit impulse (typically denoted by $\delta(t)$), the output is a rectangular pulse whose width is t_s, which can be expressed as the difference between a unit step (denoted by $u(t)$) and a delayed unit step (denoted by $u(t - t_s)$).

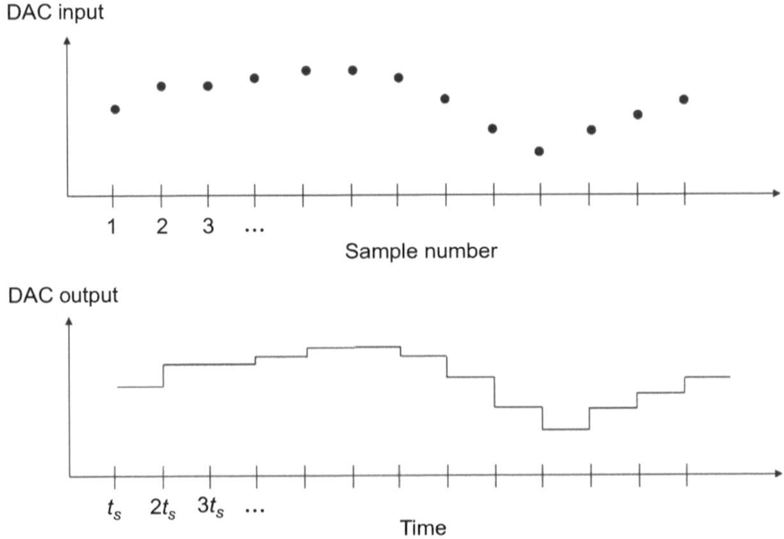

Figure 2.36 The input and output signal for a digital-to-analog converter (DAC).

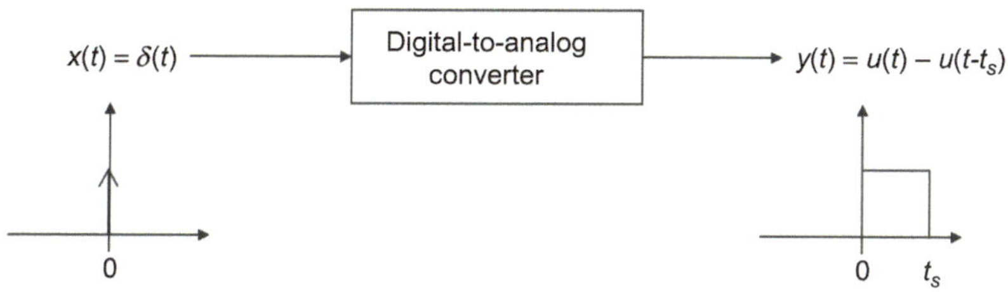

Figure 2.37 The model used to derive the transfer function of the digital-to-analog converter.

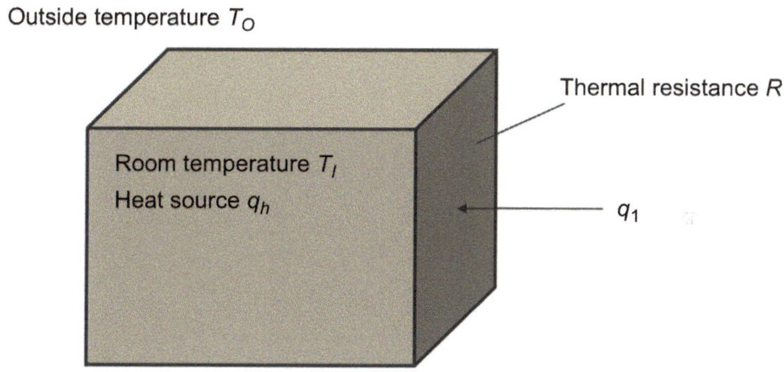

Figure 2.38 The model of heat flow for a room with perfect thermal insulation except for one wall with thermal resistance R.

See Figure 2.37. Using the fact that the Laplace transforms of $\delta(t)$, $u(t)$, and $u(t - t_s)$ are 1, $\frac{1}{s}$, and $\frac{1}{s}e^{-t_s s}$ respectively, the transfer function of the DAC is

$$\frac{Y(s)}{X(s)} = \frac{1}{s}(1 - e^{-t_s s}) \tag{2.58}$$

Often it is convenient to combine the DAC with the continuous-time plant that follows it into a single transfer function using the z-transform. We will do this in the context of the temperature control system introduced in this section. First, we derive the transfer function for the continuous element in the system. Consider a room as shown in Figure 2.38 that has perfect thermal insulation everywhere except for one wall.[6] We can derive the room's dynamical model using the governing

[6] The model here is adopted from Franklin, Powell, and Emami-Naeini (2010).

relationships for heat flow, temperature, and temperature difference. The first relationship is

$$q = \frac{1}{R}(T_1 - T_2) \tag{2.59}$$

where q is the heat flow through the material, R is its thermal resistance, and T_1 and T_2 are the temperatures on either side of the material. The second relationship is

$$\dot{T} = \frac{1}{C}q \tag{2.60}$$

where T is the temperature of an object, C is its thermal capacity, and q is the total heat flow into the object.

Applying (2.59) and (2.60) to the room, we get

$$\dot{T}_I = \frac{1}{C}(q_1 + q_h) \tag{2.61}$$

and

$$q_1 = \frac{1}{R}(T_O - T_I) \tag{2.62}$$

where T_I is the room temperature, T_O is the outside temperature, C is the thermal capacity of the room, R is the thermal resistance of the uninsulated wall, q_1 is the heat flow into the room through the uninsulated wall, and q_h is the heat source in the room.

Substituting (2.62) into (2.61), the model for the room is

$$\dot{T}_I = \frac{1}{C}\left(\frac{1}{R}(T_O - T_I) + q_h\right) \tag{2.63}$$

Simplifying the model to contain only one variable, let T_d denote the difference between the inside and outside temperatures.

$$T_d = T_I - T_O \tag{2.64}$$

Then

$$\begin{aligned} \dot{T}_d &= \dot{T}_I - \dot{T}_O \\ &= \dot{T}_I \end{aligned} \tag{2.65}$$

The last simplification can be made if we assume the outside temperature is slowly time varying compared with the inside temperature

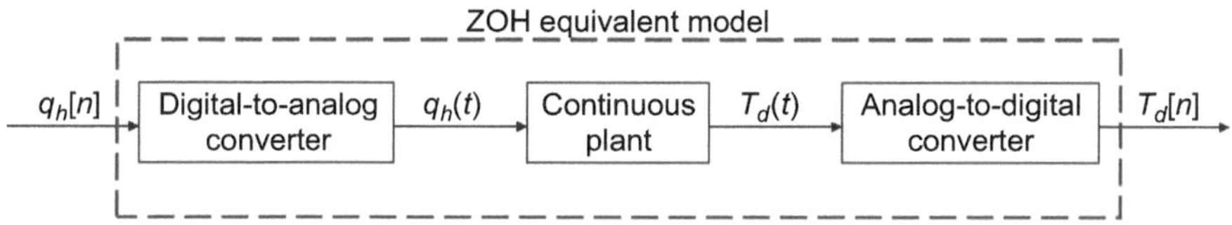

Figure 2.39 The digital-to-analog converter, continuous plant, and analog-to-digital converter can be combined to give the zero-order hold (ZOH) equivalent model.

($\dot{T}_O \approx 0$). Substituting (2.64) and (2.65) into (2.63) gives a modified model of the room.

$$\dot{T}_d = -\frac{T_d}{RC} + \frac{q_h}{C} \tag{2.66}$$

Taking the Laplace transform (because we are still dealing with a continuous-time system at this point) and rearranging gives the transfer function for the room, the continuous plant in the heating system. Denote this transfer function by $G_p(s)$.

$$G_p(s) = \frac{\overline{T}_d(s)}{Q_h(s)} = \frac{\frac{1}{C}}{s + \frac{1}{RC}} \tag{2.67}$$

Now let us combine this continuous plant with the DAC and ADC to give us a ZOH model of the plant as shown in Figure 2.39.

Working through the block diagram in Figure 2.39, we will derive the transfer function for the discrete-time system from $q_h[n]$ to $T_d[n]$. Let the input be an impulse.

$$q_n[n] = \delta[n] \tag{2.68}$$

Then taking the z-transform gives

$$Q_h(z) = 1 \tag{2.69}$$

Given the impulse input, the output of the DAC is

$$q_h(t) = u(t) - u(t - t_s) \tag{2.70}$$

where

$$u(t) = \begin{cases} 1, & t > 0 \\ 0, & t < 0 \end{cases} \tag{2.71}$$

With $u(t) - u(t - t_s)$ as the input to the plant, the output is the step response given by $T_d(t) = w(t) - w(t - t_s)$, where

$$w(t) = \mathscr{L}^{-1}\left\{\frac{G_p(s)}{s}\right\} \tag{2.72}$$

and

$$w(t - t_s) = \mathscr{L}^{-1}\left\{\frac{e^{-t_s s} G_p(s)}{s}\right\} \tag{2.73}$$

In these expressions, $\mathscr{L}^{-1}$ denotes the inverse Laplace transform. Then the output of the ADC is

$$\begin{aligned} T_d(nt_s) &= w(nt_s) - w(nt_s - t_s) \\ &= w(nt_s) - w((n-1)t_s) \\ &= w[n] - w[n-1] \end{aligned} \tag{2.74}$$

and the z-transform of the output is

$$\begin{aligned} \overline{T}_d(z) &= W(z) - z^{-1}W(z) \\ &= (1 - z^{-1})W(z) \\ &= (1 - z^{-1})Z\{w[n]\} \\ &= (1 - z^{-1})Z\{w(nt_s)\} \\ &= (1 - z^{-1})Z\left\{\mathscr{L}^{-1}\left\{\frac{G_p(s)}{s}\right\}\right\} \end{aligned} \tag{2.75}$$

In these expressions, $Z\{\cdot\}$ denotes the z-transform. The transfer function of the DAC, continuous plant, ADC series combination $G_{eq}(z)$ is then

$$G_{eq}(z) = \frac{\overline{T}_d(z)}{Q_h(z)} = (1 - z^{-1})Z\left\{\mathscr{L}^{-1}\left\{\frac{G_p(s)}{s}\right\}\right\} \tag{2.76}$$

We will now substitute the room model into (2.76).

$$\frac{G_p(s)}{s} = \frac{\frac{1}{C}}{s\left(s + \frac{1}{RC}\right)} \tag{2.77}$$

Then taking the inverse Laplace transform (obtained from a table of Laplace transforms)

$$\mathscr{L}^{-1}\left\{\frac{G_p(s)}{s}\right\} = R\left(1 - e^{-t/RC}\right) \tag{2.78}$$

and now taking the z-transform of the sampled version of this signal gives

$$
\begin{aligned}
Z\left\{\mathscr{L}^{-1}\left\{\frac{G_p(s)}{s}\right\}\right\} &= R\left(\frac{z}{z-1} - \frac{z}{z-e^{-t_s/RC}}\right) \\
&= R\,\frac{\left(1-e^{-t_s/RC}\right)z}{(z-1)\left(z-e^{-t_s/RC}\right)} \\
&= R\,\frac{1-e^{-t_s/RC}}{(1-z^{-1})\left(z-e^{-t_s/RC}\right)}
\end{aligned}
\tag{2.79}
$$

where this expression uses (2.49) and

$$
\begin{aligned}
Z(1) &= \sum_{n=0}^{\infty} z^{-n} \\
&= \frac{1}{1-z^{-1}} \\
&= \frac{z}{z-1}
\end{aligned}
\tag{2.80}
$$

Therefore, the ZOH equivalent transfer function becomes

$$
G_{eq}(z) = (1-z^{-1})R\,\frac{1-e^{-t_s/RC}}{(1-z^{-1})\left(z-e^{-t_s/RC}\right)} = R\left(\frac{1-e^{-t_s/RC}}{z-e^{-t_s/RC}}\right)
\tag{2.81}
$$

or in terms of negative powers of z,

$$
G_{eq}(z) = R\left(\frac{\left(1-e^{-t_s/RC}\right)z^{-1}}{1-e^{-t_s/RC}z^{-1}}\right)
\tag{2.82}
$$

It is worth noting here that an assumption has been made: the same sample rate t_s is used for both the analog-to-digital and digital-to-analog conversion so that the whole system is in sync. It is important to know this sampling rate because it greatly affects the system performance. Also, in this sampling process, information between samples is lost and cannot be recovered. We will see the effects of the sampling rate in the following MATLAB example.

2.3.3.1 MATLAB Example: Model of a Computer-Controlled Heating System

We will now simulate the step response of the system in (2.81). But before simulating in MATLAB, we need to establish numbers to use for R and C, and this requires a few steps.

First, there are some values that we will assume for this situation: The MATLAB code shown below provides the step response for both the continuous-time and discrete-time transfer functions of the system for side by side comparison.

```
% heating.m

close all
clear all

% Define constants for the system
ts = 10;        % sampling interval in seconds
R = 0.0242;     % Ks/J
C = 44100;      % J/K
a = exp(-ts/(R*C));

% Transfer function of continuous system
num = 1/C;
den = [1, 1/(R*C)];
Gp = tf(num,den)
subplot(2,1,1)
step(Gp,7000)
title('Step Response of Continuous System')

% Transfer function of discrete system
num = R*(1-a);
den = [1, -a];
Geq = tf(num,den,ts)
subplot(2,1,2)
step(Geq,7000)
title(['Step Response of Discrete System with Sampling Time of
',num2str(ts),' seconds'])
```

In the first part of the code, the constants in the system R and C are defined as those calculated in Table 2.5. The sampling time `ts` is initially set to 10 s, and the variable `a` is defined to simplify entering the transfer function.

The next section of code defines the continuous transfer function from (2.67). The method used here is exactly that which we have seen in Section 2.3.2.1.

The last section of code defines the discrete transfer function from (2.81). Here we see the `tf` command being used similar to the continuous case. The numerator and denominator of the transfer function must be defined by the coefficients of z in decreasing powers. In this case, positive powers of z

Table 2.5 Parameters for the Heating System

Heat capacity at constant volume of air at 300 K	$c_v = 1000$ J/(kg K)
Density of air	$\rho = 1.225$ kg/m^3
Room volume	$V = 3$ m $\times$ 4 m $\times$ 3 m $= 36$ m^3
Mass of air	$m_a = \rho V = 44.1$ kg
Thermal capacity of air in room	$\mathbf{C = c_v m_a = 44100}$ **J/K**
Thermal resistance of a brick wall	$R_B = 0.29$ m^2 K s/J
Area of brick wall	$A = 12$ m^2
Absolute thermal resistance of the brick wall	$\mathbf{R = R_B/A = 0.0242}$ **K s/J**

are used, and this is the default setting for MATLAB. The only difference between the definitions of the continuous-time and discrete-time systems is that the user must specify the sampling time.

```
Geq = tf(num,den,ts)
```

The plot resulting from this code is shown in Figure 2.40.

What do these plots tell us? Remembering back to the transfer functions in (2.67) and (2.76), they are defined as $\frac{\overline{T}_d(s)}{Q_h(s)}$ and $\frac{\overline{T}_d(z)}{Q_h(z)}$, respectively. That is, temperature over heat in both cases. The `step` command assumes the input

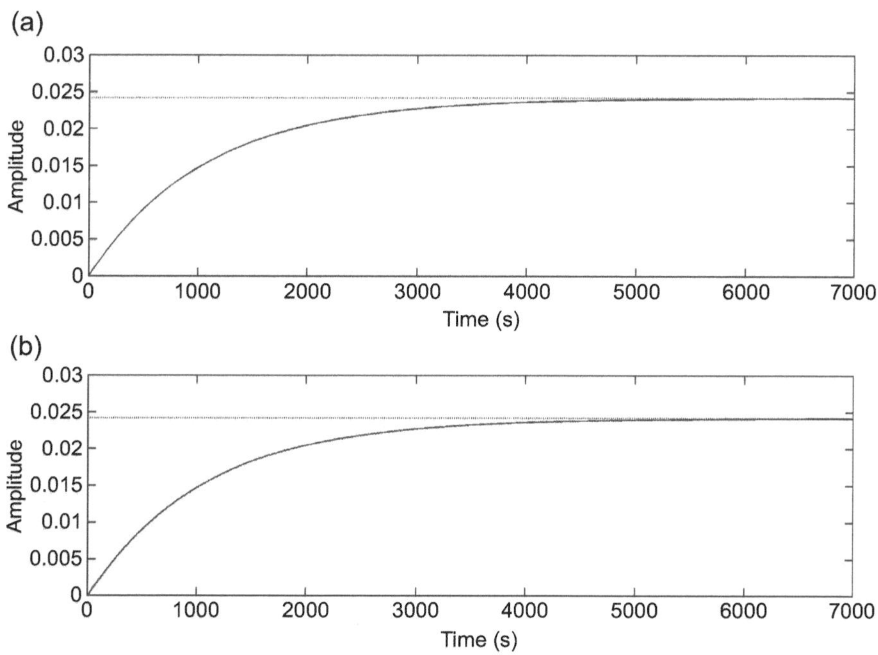

Figure 2.40 A comparison of the step response for a continuous-time system (a) and its discrete-time counterpart (b).

to the system goes from 0 to 1 at time zero, and in our case, the units of heat are J/s based on the constants we used. It is worth performing a dimensional analysis to verify what units we are dealing with and that they are the same in both systems.

Looking at the continuous case,

$$\frac{\overline{T}_d(s)}{Q_h(s)} = \frac{\frac{1}{C}}{s + \frac{1}{RC}}$$

$$= \frac{\frac{1}{[\mathrm{J/K}]}}{\frac{1}{[\mathrm{Ks/J}]\,[\mathrm{J/K}]}}$$

$$= \frac{\mathrm{Ks}}{\mathrm{J}} \tag{2.83}$$

As a result of applying heat to the system in J/s, the output is in K. Remembering the simplification we performed in the model derivation, T_d represents the difference between the room temperature and the (constant) outside temperature. Therefore, the step response plot shows us that when the system has a heat input of 1 J/s, the temperature difference increases from 0 to approximately 0.025 K in 6500 s.

We see the shape of a first-order system as we expect because of the highest power of s (or z) being 1 in the transfer function with no overshoot or oscillation. Because the system is linear, if we wanted to know the response to an input of 50 J/s or 0.025 J/s, we simply scale the output by 50 or 0.025, respectively.

Using the above code, it is relatively easy to see the effects of sampling time on the output. In the above figure, the sampling time is 10 s. For this sampling time the continuous and discrete responses look the same. The reason for this is that 10 s is fast compared to the overall response time of the system. It takes the system thousands of seconds (over an hour) to reach its final temperature. Changing the sampling time to 120 s (or 2 min) yields the plot shown in Figure 2.41. In this case, we see the effects of sampling

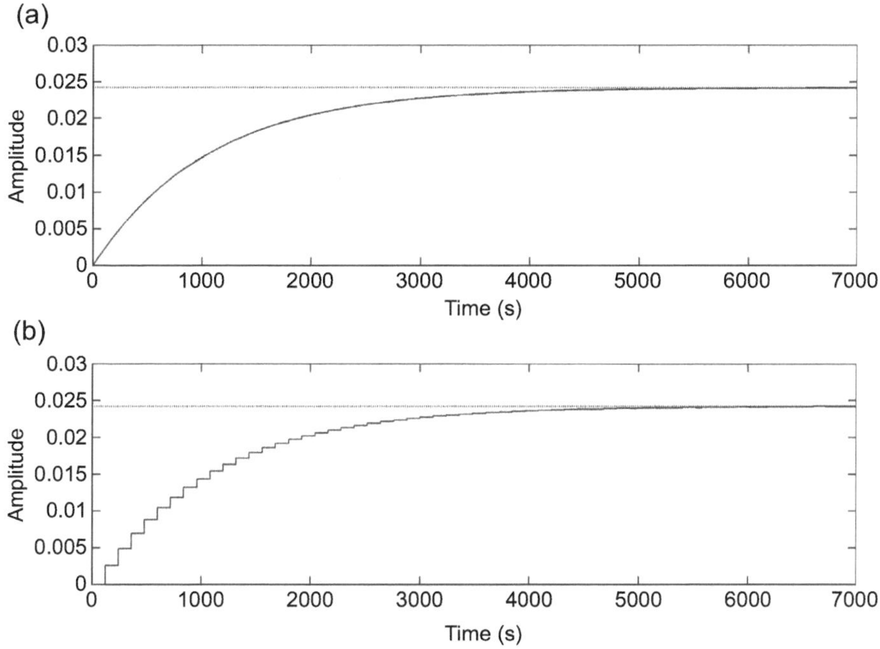

Figure 2.41 A comparison of the step response for a continuous-time system (a) and its discrete-time counterpart (b) with a sampling time of 120 s.

in the discrete-time system as it shows a stepwise increase in its response, but the final value of the amplitude remains the same as its continuous counterpart.

2.4 STATE-SPACE REPRESENTATION

2.4.1 Overview

So far in this chapter, we have represented systems using their differential or difference equations and transfer functions. We saw how transfer functions can be derived from equations of motion and how convenient they were to use in MATLAB. In this section, we discuss an alternative method: state-space representation. Using this method, we rearrange the equations of motion into a specific format using matrices and the state variables of the system. It is here where we will see that an understanding of linear algebra will be of great use.

One of the defining characteristics of transfer functions is that they can only be used to represent linear systems with zero initial conditions. We saw in the previous example that the transfer function method breaks down when we try to incorporate nonzero initial conditions. Another characteristic of transfer functions is that they give only the input—output relationship. For example, in the car suspension model, the transfer function was $\frac{X_2(s)}{R(s)}$, giving us the relationship between the road elevation (input) and the car body position (output). Although this is ultimately the relationship that interests us, it gives no indication of the wheel height x_1, or time-changing behavior of either the car body or wheel, $\dot{x}_1, \dot{x}_2$. **State-space models** overcome these weaknesses and give us a fuller picture of the system.

The major characteristics of state-space representation are

- It provides modeling for linear *and* nonlinear systems and thus can model real-world systems more accurately than transfer functions.
- It gives access to internal behavior and not only input—output behavior.
- There are several well-known standard forms, and many techniques have been developed to design controllers based on these standard forms.
- It easily models multiple input, multiple output (MIMO) systems.
- A state-space model of a given system is not unique; many equivalent models exist, and one is free to choose the one that is most convenient or useful for a given circumstance.

The general form for a linear state-space model is

$$\begin{aligned} \dot{x}(t) &= Ax(t) + Bu(t) \\ y(t) &= Cx(t) + Du(t) \end{aligned} \tag{2.84}$$

for continuous-time systems or

$$\begin{aligned} x[n+1] &= Ax[n] + Bu[n] \\ y[n] &= Cx[n] + Du[n] \end{aligned} \tag{2.85}$$

for discrete-time systems.

In both (2.84) and (2.85), the first equation is called the **state equation,** and the second equation is called the **output equation**. Also, the conventions in Table 2.6 are used.

For single-input, single-output (SISO) systems, the B and C matrices reduce to vectors of size Nx1, and 1xN, respectively, and are denoted by b and c. Also, the D matrix reduces to a scalar d.

Also, the models in (2.84) and (2.85) are for a time-invariant system. For time-varying systems, the matrix elements become functions of time $A(t)$, $B(t)$, $C(t)$, and $D(t)$.

Table 2.6 Definitions of the State-Space Matrices

$x = \begin{bmatrix} x_1 \\ x_2 \\ \vdots \\ x_N \end{bmatrix}$	State vector: a vector containing the N states of the Nth order system*
$u = \begin{bmatrix} u_1 \\ u_2 \\ \vdots \\ u_p \end{bmatrix}$	Input vector: a vector containing the p inputs of the system
$y = \begin{bmatrix} y_1 \\ y_2 \\ \vdots \\ y_q \end{bmatrix}$	Output vector: a vector containing the q outputs of the system
$A = \begin{bmatrix} a_{11} & a_{12} & \cdots & a_{1N} \\ a_{21} & a_{22} & \cdots & a_{2N} \\ \vdots & \vdots & \ddots & \vdots \\ a_{N1} & a_{N2} & \cdots & a_{NN} \end{bmatrix}$	State matrix: an $N \times N$ matrix relating the states to their derivatives (continuous) or next value (discrete)
$B = \begin{bmatrix} b_{11} & b_{12} & \cdots & b_{1p} \\ b_{21} & b_{22} & \cdots & b_{2p} \\ \vdots & \vdots & \ddots & \vdots \\ b_{N1} & b_{N2} & \cdots & b_{Np} \end{bmatrix}$	Input matrix: an $N \times p$ matrix relating the inputs to the derivatives of the states
$C = \begin{bmatrix} c_{11} & c_{12} & \cdots & c_{1N} \\ c_{21} & c_{22} & \cdots & c_{2N} \\ \vdots & \vdots & \ddots & \vdots \\ c_{q1} & c_{q2} & \cdots & c_{qN} \end{bmatrix}$	Output matrix: a $q \times N$ matrix relating the states to the outputs
$D = \begin{bmatrix} d_{11} & d_{12} & \cdots & d_{1p} \\ d_{21} & d_{22} & \cdots & d_{2p} \\ \vdots & \vdots & \ddots & \vdots \\ d_{q1} & d_{q2} & \cdots & d_{qp} \end{bmatrix}$	Feedthrough matrix: a $q \times p$ matrix relating the input and output (so called because it indicates how the input feeds directly through the system to the output)

*Although it may be more typical to denote the order of a system by n, the upper case N is used to avoid confusion with the discrete-time variable.

2.4.2 What Is a State?

So far in this section, the term "state" has been used several times without definition. State can mean many things depending on the circumstances; some familiar notions of state include governmental communities (e.g., Rhode Island) in a political context and solid, liquid, or gas in a chemistry context. For our purposes in discussing dynamical systems, the state refers to an abstract notion of where the system is located in a geometrical space.

A solid, useful definition of *state* is hard to come by. Many books simply introduce the notion through the dynamical equations of the system and call the variables in the equations the "state variables" because they describe the state of the system. Such introductions can be frustrating because they are vague and do not indicate where they come from or how to choose them. The vagueness is exactly what makes state-space modeling so powerful. We can choose the states however it is most appropriate for us. And we can transform one model into another one that is equivalent but in a more usable form. The state of a system is unique, but how we represent it is not.

It is often useful to start with a physical system in which the states are straightforward. One useful explanation of states comes from Cannon (2003). He explains:

> *What we mean by the state of a mechanical system may be introduced by analogy with configuration: The state of a system is often defined by an independent set of* position coordinates, plus *their* derivatives. *Thus, state implies* configuration plus velocity. *Configuration tells only where the system is, but state tells us both where it is and how fast (and in what direction) it is going; and as expressed by Professor L. Sadeh, "the state of a system at a given time, plus its differential equations of motion and inputs, will determine its configuration for all future time."*

Let's see how this relates to the car suspension system, whose equations of motion are repeated below. There are two masses in the system and thus two position coordinates x_1 and x_2 giving us the system configuration. These coordinates, together with their derivatives $\dot{x}_1$ and $\dot{x}_2$, give the complete information about how the system will evolve from a given starting point (the initial state $x_1(0), \dot{x}_1(0), x_2(0), \dot{x}_2(0)$)

$$\begin{aligned}
\ddot{x}_1(t) + \frac{b}{m_1}\dot{x}_1(t) + \frac{k_1+k_2}{m_1}x_1(t) - \frac{b}{m_1}\dot{x}_2(t) - \frac{k_2}{m_1}x_2(t) &= \frac{k_1}{m_1}r(t) \\
\ddot{x}_2(t) + \frac{b}{m_2}\dot{x}_2(t) + \frac{k_2}{m_2}x_2(t) - \frac{b}{m_2}\dot{x}_1(t) - \frac{k_2}{m_2}x_1(t) &= 0
\end{aligned} \tag{2.86}$$

Table 2.7 Explanation of the State Variables

State Symbol in (2.88)	New State Symbol to Use in (2.84)	Variable in Original Equations (2.86)	Physical Meaning
y_1	x_1	$\dot{x}_1$	Wheel height rate of change
y_2	x_2	x_1	Wheel height
y_3	x_3	$\dot{x}_2$	Car body height rate of change
y_4	x_4	x_2	Car body height

The state-space model for this system was derived in MATLAB Example 2.2.1.1 and was given as

$$\begin{aligned}
\dot{y}_1 &= -\frac{b}{m_1}y_1 - \frac{k_1+k_2}{m_1}y_2 + \frac{b}{m_1}y_3 + \frac{k_2}{m_1}y_4 + \frac{k_1}{m_1}r \\
\dot{y}_2 &= y_1 \\
\dot{y}_3 &= -\frac{b}{m_2}y_3 - \frac{k_2}{m_2}y_4 + \frac{b}{m_2}y_1 + \frac{k_2}{m_2}y_2 \\
y_4 &= y_3
\end{aligned} \tag{2.87}$$

where

$$\begin{aligned}
y_1 &= \dot{x}_1 \\
y_2 &= x_1 \\
y_3 &= \dot{x}_2 \\
y_4 &= x_2
\end{aligned} \tag{2.88}$$

Now to put this system model into the form given by (2.84), we will do some rearranging and possibly add confusion because of multiple meanings of the symbol x. These conversions are given in Table 2.7.

Two more changes need to be made. We will also choose an output y. In this case, the car body height, now known as x_4, will be used. Finally, the input r to the system will be denoted by u.

Rewriting (2.87) with the new variable definitions gives

$$\begin{aligned}
\dot{x}_1 &= -\frac{b}{m_1}x_1 - \frac{k_1+k_2}{m_1}x_2 + \frac{b}{m_1}x_3 + \frac{k_2}{m_1}x_4 + \frac{k_1}{m_1}u \\
\dot{x}_2 &= x_1 \\
\dot{x}_3 &= -\frac{b}{m_2}x_3 - \frac{k_2}{m_2}x_4 + \frac{b}{m_2}x_1 + \frac{k_2}{m_2}x_2 \\
\dot{x}_4 &= x_3
\end{aligned} \tag{2.89}$$

Rearranging into matrix form gives the state-space model for the system.

$$\begin{bmatrix} \dot{x}_1 \\ \dot{x}_2 \\ \dot{x}_3 \\ \dot{x}_4 \end{bmatrix} = \begin{bmatrix} -\frac{b}{m_1} & -\frac{k_1+k_2}{m_1} & \frac{b}{m_1} & \frac{k_2}{m_1} \\ 1 & 0 & 0 & 0 \\ \frac{b}{m_2} & \frac{k_2}{m_2} & -\frac{b}{m_2} & -\frac{k_2}{m_2} \\ 0 & 0 & 1 & 0 \end{bmatrix} \begin{bmatrix} x_1 \\ x_2 \\ x_3 \\ x_4 \end{bmatrix} + \begin{bmatrix} \frac{k_1}{m_1} \\ 0 \\ 0 \\ 0 \end{bmatrix} u \tag{2.90}$$

$$y = \begin{bmatrix} 0 & 0 & 0 & 1 \end{bmatrix} \begin{bmatrix} x_1 \\ x_2 \\ x_3 \\ x_4 \end{bmatrix} + [0]u$$

Thus, the matrices are

$$A = \begin{bmatrix} -\frac{b}{m_1} & -\frac{k_1+k_2}{m_1} & \frac{b}{m_1} & \frac{k_2}{m_1} \\ 1 & 0 & 0 & 0 \\ \frac{b}{m_2} & \frac{k_2}{m_2} & -\frac{b}{m_2} & -\frac{k_2}{m_2} \\ 0 & 0 & 1 & 0 \end{bmatrix} \qquad b = \begin{bmatrix} \frac{k_1}{m_1} \\ 0 \\ 0 \\ 0 \end{bmatrix}$$

$$c = \begin{bmatrix} 0 & 0 & 0 & 1 \end{bmatrix} \qquad d = [0]$$

2.4.2.1 MATLAB Example: State-Space Model of the Car Suspension

The MATLAB code below shows how to enter the state-space model (2.90) using the `ss` command.

```
% statespace.m

close all
clear all

% Define the model parameters
m1 = 70;           % kg
m2 = 350;          % kg
k1 = 176000;       % N/m
k2 = 27000;        % N/m
b = 2500;          % Ns/m

% Define the A,B,C,D matrices
A = [-b/m1, -(k1+k2)/m1, b/m1, k2/m1; 1, 0, 0, 0; b/m2, k2/m2, -b/m2,
-k2/m2; 0, 0, 1, 0];
B = [k1/m1; 0; 0; 0];
C = [0, 0, 0, 1];
D = [0];

sys = ss(A,B,C,D)
step(sys)
```

Entering the state-space model is as easy, if not easier, than entering the transfer function. One simply has to directly create the A, B, C, and D matrices, separating elements in a row by a comma[7] and separating rows by semicolons. Then the `ss` command is used with those matrices as arguments. In the code above, the line defining the system does not have a semicolon to suppress the output, so the command window shows the following information when the simulation is run.

```
sys =

  a =
             x1        x2        x3        x4
   x1    -35.71     -2900     35.71     385.7
   x2         1         0         0         0
   x3     7.143     77.14    -7.143    -77.14
   x4         0         0         1         0
```

[7] In MATLAB, row elements may be separated by either a comma or a space. Commas are used here to make the element separation more clear.

```
b =
          u1
   x1   2514
   x2      0
   x3      0
   x4      0

c =
       x1 x2 x3 x4
   y1   0  0  0  1

d =
       u1
   y1   0

Continuous-time state space model.
```

Checking the result is always a good idea to make sure the matrices were entered correctly.

The step response plot resulting from the `step` command is shown in Figure 2.42. It is identical to the one obtained in MATLAB Example 2.3.2.1 shown in Figure 2.26.

One powerful and relatively simple thing we can do with the state-space model in MATLAB is to plot the step response of any state. If we wanted to

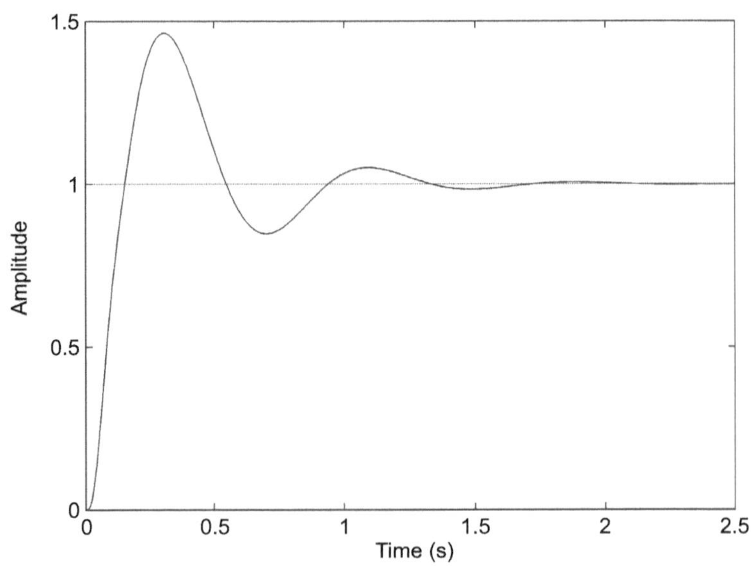

Figure 2.42 The step response of the car suspension obtained from the state-space model.

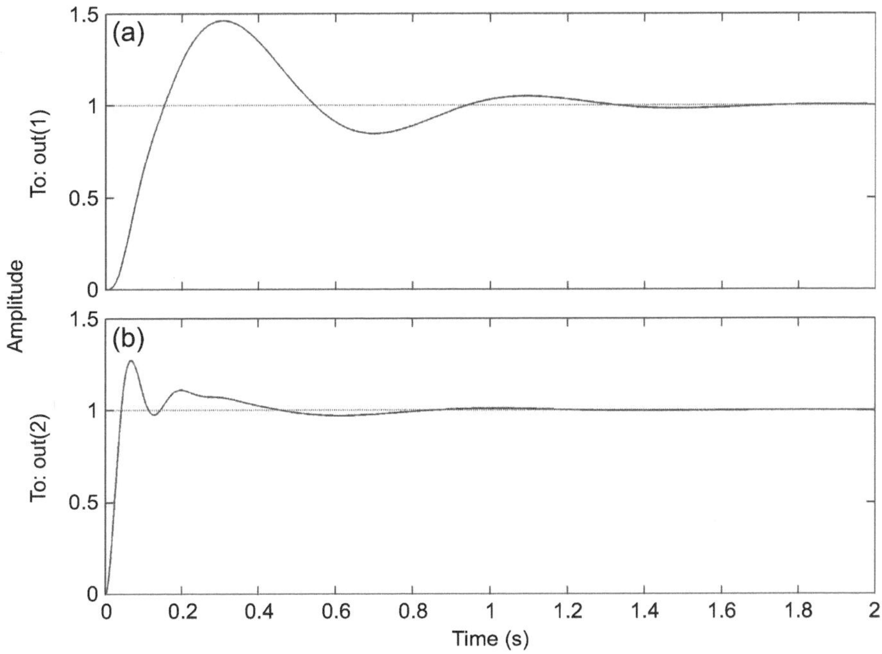

Figure 2.43 The step response of the car suspension showing two states, the car body height ×4 (a) and the wheel height ×2 (b).

do this with transfer functions, we would need to find an expression relating the input to each output. But with the state-space model, we only have to modify the C matrix to include whichever state we want as an output.

Suppose we wish to plot the step response for not only the car body height but also the wheel height. To do this, we modify the C matrix to be

$$C = \begin{bmatrix} 0 & 0 & 0 & 1 \\ 0 & 1 & 0 & 0 \end{bmatrix}$$

Or in MATLAB code

```
C = [0, 0, 0, 1; 0, 1, 0, 0];
```

In this case, there are two outputs: the first one corresponding to x_4, the car body height, and the second one corresponding to x_2, the wheel height. The resulting step response plot is shown in Figure 2.43.

The state-space model in (2.90) is not unique. In fact, there are infinitely many models one could develop, although they would not all be useful. One obvious way to get another representation is to rearrange the order of the states. For example, we could have let x_1 be the car body height, x_2 be the car body height rate of change, and so on. In this case, the elements of the matrices would be the same, but rows or columns would be switched. Another way is to use a scale factor such as letting x_1 be twice the car height. Although this latter method is mathematically correct, it may not make much sense physically.

One way to get another model that may make sense physically is to choose one of the states to be the difference between the wheel height and car body height. Perhaps there is a sensor on the car that allows us to measure this quantity directly, but we cannot measure the wheel height. In this scenario, redefine the states and denote them by $\widehat{x}$ as follows.

$$\begin{aligned}\widehat{x}_1 &= x_3 - x_1\\ \widehat{x}_2 &= x_4 - x_2\\ \widehat{x}_3 &= x_3\\ \widehat{x}_4 &= x_4\end{aligned} \tag{2.91}$$

Substituting in from (2.89), using $x_1 = \widehat{x}_3 - \widehat{x}_1$ and $x_2 = \widehat{x}_4 - \widehat{x}_2$, and simplifying gives

$$\begin{aligned}\dot{\widehat{x}}_1 &= \dot{x}_3 - \dot{x}_1\\ &= -\frac{b}{m_2}x_3 - \frac{k_2}{m_2}x_4 + \frac{b}{m_2}x_1 + \frac{k_2}{m_2}x_2\\ &\quad - \left(-\frac{b}{m_1}x_1 - \frac{k_1+k_2}{m_1}x_2 + \frac{b}{m_1}x_3 + \frac{k_2}{m_1}x_4 + \frac{k_1}{m_1}u\right)\\ &= -\left(\frac{b}{m_1} + \frac{b}{m_2}\right)\widehat{x}_1 - \left(\frac{k_1+k_2}{m_1} + \frac{k_2}{m_2}\right)\widehat{x}_2 + \frac{k_1}{m_1}\widehat{x}_4 - \frac{k_1}{m_1}u\end{aligned} \tag{2.92}$$

$$\begin{aligned}\dot{\widehat{x}}_2 &= \dot{x}_4 - \dot{x}_2\\ &= x_3 - x_1\\ &= \widehat{x}_1\end{aligned} \tag{2.93}$$

$$\begin{aligned}\dot{\widehat{x}}_3 &= \dot{x}_3\\ &= -\frac{b}{m_2}x_3 - \frac{k_2}{m_2}x_4 + \frac{b}{m_2}x_1 + \frac{k_2}{m_2}x_2\\ &= -\frac{b}{m_2}\widehat{x}_1 - \frac{k_2}{m_2}\widehat{x}_2\end{aligned} \tag{2.94}$$

$$\begin{aligned}\dot{\widehat{x}}_4 &= \dot{x}_4\\ &= x_3\\ &= \widehat{x}_3\end{aligned} \tag{2.95}$$

Finally, the state-space model becomes

$$\begin{bmatrix} \dot{\hat{x}}_1 \\ \dot{\hat{x}}_2 \\ \dot{\hat{x}}_3 \\ \dot{\hat{x}}_4 \end{bmatrix} = \begin{bmatrix} -\left(\frac{b}{m_1}+\frac{b}{m_2}\right) & -\left(\frac{k_1+k_2}{m_1}+\frac{k_2}{m_2}\right) & 0 & \frac{k_1}{m_1} \\ 1 & 0 & 0 & 0 \\ -\frac{b}{m_2} & -\frac{k_2}{m_2} & 0 & 0 \\ 0 & 0 & 1 & 0 \end{bmatrix} \begin{bmatrix} \hat{x}_1 \\ \hat{x}_2 \\ \hat{x}_3 \\ \hat{x}_4 \end{bmatrix} + \begin{bmatrix} -\frac{k_1}{m_1} \\ 0 \\ 0 \\ 0 \end{bmatrix} u \tag{2.96}$$

$$y = \begin{bmatrix} 0 & 0 & 0 & 1 \end{bmatrix} \begin{bmatrix} \hat{x}_1 \\ \hat{x}_2 \\ \hat{x}_3 \\ \hat{x}_4 \end{bmatrix} + [0]u$$

2.4.2.2 MATLAB Example: Alternate State-Space Model of the Car Suspension

To verify that the model in (2.96) is correct, we simulate the system in MATLAB with the following code.

```
% statespace_alternate.m

close all
clear all

% Define the model parameters
m1 = 70;          % kg
m2 = 350;         % kg
k1 = 176000;      % N/m
k2 = 27000;       % N/m
b = 2500;         % Ns/m

% Define the A,B,C,D matrices
A = [-(b/m1+b/m2), -((k1+k2)/m1+k2/m2), 0, k1/m1; 1, 0, 0, 0; -b/m2,
-k2/m2, 0, 0; 0, 0, 1, 0];
B = [-k1/m1; 0; 0; 0];
C = [0, 0, 0, 1;0, 1, 0, 0];
D = [0];

sys = ss(A,B,C,D)
step(sys)
```

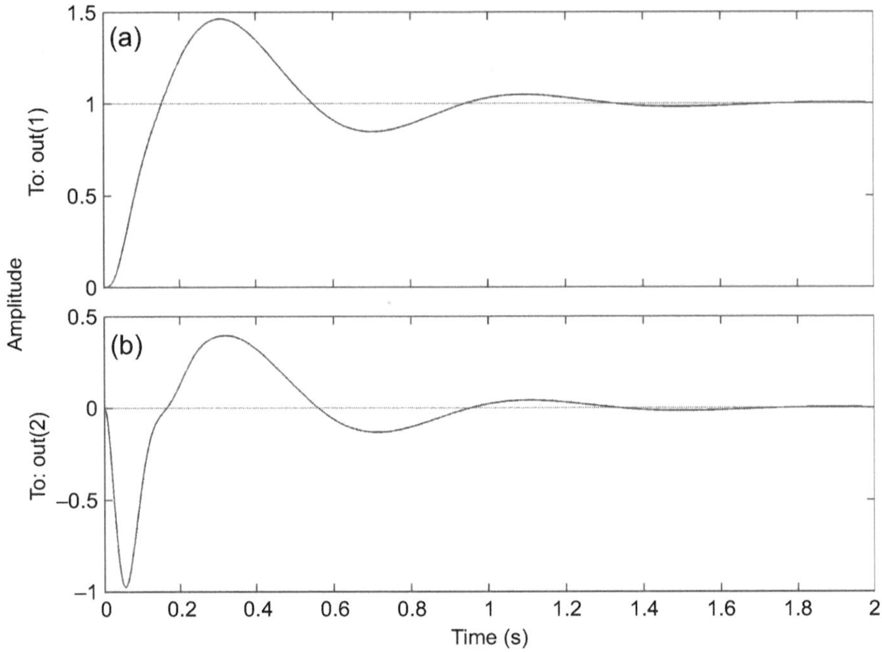

Figure 2.44 The step response of the car suspension system with redefined states (a and b).

As in the previous example, we redefine the C matrix to include both $\widehat{x}_4$ and $\widehat{x}_2$ as outputs. The resulting step response plot is shown in Figure 2.44.

The upper plot in Figure 2.44 is the first output, defined to be $\widehat{x}_4$ or the car body height. The lower plot is the second output is defined to be $\widehat{x}_2$, the body height minus the wheel height. Looking at the lower plot, notice that $\widehat{x}_2$ becomes negative after hitting the bump in the road at time zero. Does this mean that the wheel travels up and above the car's body frame? How is that possible? The answer lies in how we defined each of these heights. The zero point of each variable was its equilibrium, that is, if the car is sitting still, these values are zero. Thus, x_4 and x_2 from the original model measure how far the body and wheel has varied from these starting values. The negative value of $\widehat{x}_2$ means that the wheel has moved up from its original position more than the car body has, as we expect when the car hits a bump in the road.

How do we choose states? The car suspension model gives us a nice example for applying state-space modeling because it provides a physical example for which we have some intuition. However, we can think of these models more abstractly and still be able to choose appropriate states.

As a general rule, choose states as those things for which you can take a derivative.

Take, for example, the following system equations. Assume u and v are the inputs to the system; z_1 and z_2 are variables; and α_1, α_2, α_3, and β are constants.

$$\begin{aligned} u &= \alpha_1 z_1 + \alpha_2 \dot{z}_2 + \alpha_3 (z_1 - v) \\ z_1 - z_2 &= \beta \ddot{z}_2 \end{aligned} \tag{2.97}$$

How do we go about finding the state-space model for the system? Can we simply define the states as $x_1 = z_1$ and $x_2 = z_2$? No, because in the state-space model, we need to have an expression for the derivatives of the states in terms of the states and inputs. In this case, $\dot{z}_1$ does not show up in (2.97), so there is no expression for it. The only derivatives are of z_2, a first and second derivative. Therefore, it makes sense to define the states to be

$$\begin{aligned} x_1 &= z_2 \\ x_2 &= \dot{z}_2 \end{aligned} \tag{2.98}$$

Thus

$$\begin{aligned} \dot{x}_1 &= \dot{z}_2 \\ &= x_2 \end{aligned} \tag{2.99}$$

and

$$\begin{aligned} \dot{x}_2 &= \ddot{z}_2 \\ &= \frac{1}{\beta} z_1 - \frac{1}{\beta} z_2 \\ &= \frac{1}{\beta}\left(\frac{1}{\alpha_1 + \alpha_3} u - \frac{\alpha_2}{\alpha_1 + \alpha_3} \dot{z}_2 + \frac{\alpha_3}{\alpha_1 + \alpha_3} v\right) - \frac{1}{\beta} z_2 \\ &= -\frac{1}{\beta} x_1 - \frac{\alpha_2}{\beta(\alpha_1 + \alpha_3)} x_2 + \frac{1}{\beta(\alpha_1 + \alpha_3)} u + \frac{\alpha_3}{\beta(\alpha_1 + \alpha_3)} v \end{aligned} \tag{2.100}$$

The expressions in (2.99) and (2.100) establish the A and B matrices.

To establish c and d, we first need to know what the output of our system is. If the output is one of the states, then $d = 0$ and $c = \begin{bmatrix} 1 & 0 \end{bmatrix}$ or $c = \begin{bmatrix} 0 & 1 \end{bmatrix}$, depending on whether it is the first or second state. However, the situation may not be so simple, and the output may be some combination of all states and inputs. For example, suppose the variables z_1 and z_2 represent voltages in a circuit and we wish to take the output to be one of the currents, say $\alpha_1 z_1$. So,

$$y = \alpha_1 z_1 \tag{2.101}$$

How do we express y as a function of x_1, x_2, u, and v? Or equivalently, how do we express y as a function of z_2, $\dot{z}_2$, u, and v? As a first attempt, let us use the second equation in (2.97) and solve for z_1.

$$z_1 = \beta \ddot{z}_2 + z_2 \tag{2.102}$$

Then y becomes

$$y = \alpha_1 (\beta \ddot{z}_2 + z_2) \tag{2.103}$$

However, $\ddot{z}_2$ is not a state of the system; it is a derivative of a state. We cannot use (2.97) to express $\ddot{z}_2$ as a function of the states and inputs and perform a substitution. Therefore, this is a path that will not lead us to what we need.

Alternatively, let's start with the first equation in (2.97), solve for z_1, then multiply the result by α_1.

$$\begin{aligned} z_1 &= -\frac{\alpha_2}{\alpha_1 + \alpha_3}\dot{z}_2 + \frac{1}{\alpha_1 + \alpha_3}u + \frac{\alpha_3}{\alpha_1 + \alpha_3}v \\ \alpha_1 z_1 &= -\frac{\alpha_1\alpha_2}{\alpha_1 + \alpha_3}\dot{z}_2 + \frac{\alpha_1}{\alpha_1 + \alpha_3}u + \frac{\alpha_1\alpha_3}{\alpha_1 + \alpha_3}v \end{aligned} \tag{2.104}$$

Therefore, the expression for y becomes a function of the states and inputs.

$$y = -\frac{\alpha_1\alpha_2}{\alpha_1 + \alpha_3}x_2 + \frac{\alpha_1}{\alpha_1 + \alpha_3}u + \frac{\alpha_1\alpha_3}{\alpha_1 + \alpha_3}v \tag{2.105}$$

Finally, using (2.99), (2.100), and (2.105), the state-space model for the system is

$$\begin{aligned} \begin{bmatrix} \dot{x}_1 \\ \dot{x}_2 \end{bmatrix} &= \begin{bmatrix} 0 & 1 \\ -\dfrac{1}{\beta} & -\dfrac{\alpha_2}{(\alpha_1+\alpha_3)\beta} \end{bmatrix} \begin{bmatrix} x_1 \\ x_2 \end{bmatrix} + \begin{bmatrix} 0 & 0 \\ \dfrac{1}{(\alpha_1+\alpha_3)\beta} & \dfrac{\alpha_3}{(\alpha_1+\alpha_3)\beta} \end{bmatrix} \begin{bmatrix} u \\ v \end{bmatrix} \\ y &= \begin{bmatrix} 0 & -\dfrac{\alpha_1\alpha_2}{\alpha_1+\alpha_3} \end{bmatrix} \begin{bmatrix} x_1 \\ x_2 \end{bmatrix} + \begin{bmatrix} \dfrac{\alpha_1}{\alpha_1+\alpha_3} & \dfrac{\alpha_1\alpha_3}{\alpha_1+\alpha_3} \end{bmatrix} \begin{bmatrix} u \\ v \end{bmatrix} \end{aligned} \tag{2.106}$$

2.4.2.3 MATLAB Example: System with Nonzero Initial Conditions

In this example, we revisit the circuit shown in Figure 2.31. In deriving the equations for the system, we saw that it was impossible to obtain a transfer function if there was an initial charge on the capacitor. Here we will show how to obtain the system response in MATLAB when there are nonzero initial conditions.

There are three basic steps to the procedure.

1. Specify the numerator and denominator of the transfer function.
2. Convert the system to its state-space form using the `tf2ss` command.
3. Use the `lsim` command and specify the initial condition as a parameter.

Notice that we are starting with the transfer function because it is given in (2.44). We could have started with the state-space model and directly gone to step 3. However, this method allows us to demonstrate how easy the conversion is from transfer functions to state-space models in MATLAB.

```
% circuit.m

close all
clear all

% Define constants for the system
R1 = 10000;       % Ohms
C1 = 0.1e-6;      % Farads

% Define constants for the simulation
vc = 2;           % Volts (initial charge of the capacitor)
t_end = 0.02;     % seconds
T = 0.0001;       % seconds
f = 100;          % Hertz

num = 1/(R1*C1);
den = [1 1/(R1*C1)];

[A B C D] = tf2ss(num,den)
sys = ss(A,B,C,D)

t = [0:T:t_end];
u = sin(2*pi*f*t);
x = R1*C1*vc;     % Scale factor
lsim(sys,u,t,x)
```

In the above code, we simulate the system with a sinusoidal input and an initial capacitor voltage of 2 V. The resulting plot is shown in Figure 2.45.

In the code, first the constants for the system and simulation are defined. Then the numerator and denominator of the transfer function are defined. The state-space model is generated in the next two lines of code using the `tf2ss` and `ss` commands.

```
[A B C D] = tf2ss(num,den)
sys = ss(A,B,C,D)
```

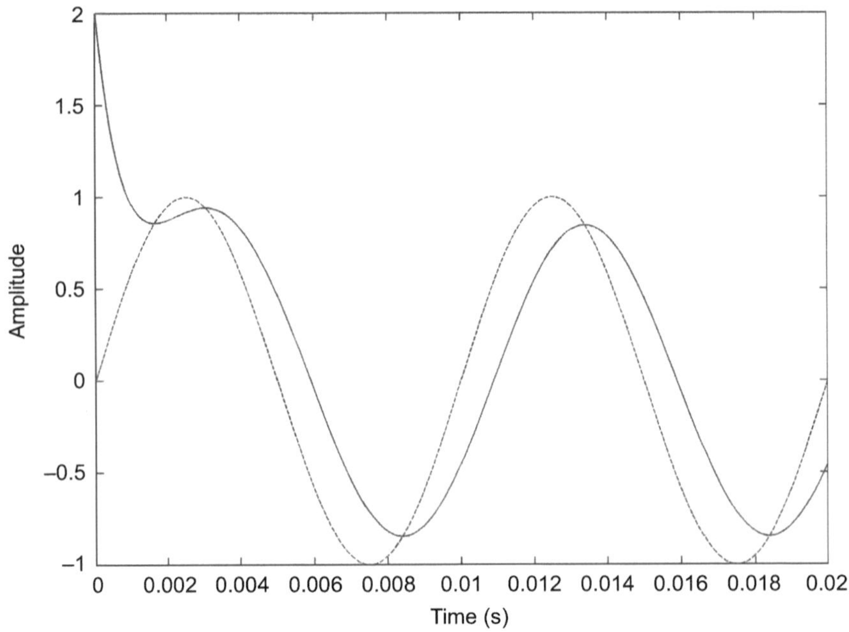

Figure 2.45 Response of the circuit with a nonzero initial condition. The circuit input is the dashed line, and the output is the solid line.

The `tf2ss` command must have two arguments specifying the numerator and denominator; thus, the method previously discussed of defining `s = tf('s')` will not work in this scenario. The output of the command is the four matrices defining the state-space model. Next the `ss` command combines these four matrices into a structure defining the system. The output of the `ss` command is

```
sys =

  a =
             x1
     x1   -1000

  b =
            u1
     x1      1
```

```
c =
        x1
  y1  1000

d =
        u1
  y1     0

Continuous-time state space model.
```

Because the system is first order, the matrices are all 1 × 1.

As discussed earlier, the state-space model for a system is not unique. In that case, how does MATLAB choose the matrices if there are infinitely many to choose from? The MATLAB designers decided to use what is known as controller canonical form. The various canonical forms will be discussed in more detail in the next section, but in this example, the transfer function

$$\frac{V_c(s)}{V_{in}(s)} = \frac{\frac{1}{RC}}{s + \frac{1}{RC}} \tag{2.107}$$

turns into the following state-space model.

$$\begin{aligned} \dot{x} &= -\frac{1}{RC}x + v_{in} \\ v_c &= \frac{1}{RC}x \end{aligned} \tag{2.108}$$

The next three lines of code define the time vector, input vector, and initial condition for use in the `lsim` command.

```
t = [0:T:t_end];
u = sin(2*pi*f*t);
x = R1*C1*vc;     % Scale factor
```

In this case, we are setting an initial condition for the capacitor voltage v_c of 2 V, but the `lsim` command requires the initial condition of the state x. Thus, the scale factor of RC must be used.

The final line, `lsim(sys,u,t,x)`, plots the system response for the given input `u`, time `t`, and initial state of the system `x`. Note that in this example, because it is a first-order system, the initial state is a single value. In general, for an Nth order system, the N initial conditions must be specified in a vector.

2.4.3 Relationship between Transfer Functions and State-space Models

The previous MATLAB example shows how to obtain a state-space model directly from the transfer function using the `tf2ss` command. Similarly, there is a command to do the reverse operation, `ss2tf`, transforming a model from state-space form to a transfer function. Here and in the next section, the methods behind these commands are discussed.

Mathematically, there is a straightforward calculation to obtain a transfer function from a state-space model.

$$\frac{Y(s)}{U(s)} = c(sI - A)^{-1}b + d \tag{2.109}$$

In this expression, I is the $N \times N$ identity matrix.

As an example, applying (2.109) to the state-space model of the car suspension (2.90) gives

$$\frac{Y(s)}{U(s)} = [0\ 0\ 0\ 1]\left(s\begin{bmatrix} 1 & 0 & 0 & 0 \\ 0 & 1 & 0 & 0 \\ 0 & 0 & 1 & 0 \\ 0 & 0 & 0 & 1 \end{bmatrix} - \begin{bmatrix} -\frac{b}{m_1} & -\frac{k_1+k_2}{m_1} & \frac{b}{m_1} & \frac{k_2}{m_1} \\ 1 & 0 & 0 & 0 \\ \frac{b}{m_2} & \frac{k_2}{m_2} & -\frac{b}{m_2} & -\frac{k_2}{m_2} \\ 0 & 0 & 1 & 0 \end{bmatrix}\right)^{-1}\begin{bmatrix} \frac{k_1}{m_1} \\ 0 \\ 0 \\ 0 \end{bmatrix}$$

$$= [0\ 0\ 0\ 1]\begin{bmatrix} s+\frac{b}{m_1} & \frac{k_1+k_2}{m_1} & -\frac{b}{m_1} & -\frac{k_2}{m_1} \\ -1 & s & 0 & 0 \\ -\frac{b}{m_2} & -\frac{k_2}{m_2} & s+\frac{b}{m_2} & \frac{k_2}{m_2} \\ 0 & 0 & -1 & s \end{bmatrix}^{-1}\begin{bmatrix} \frac{k_1}{m_1} \\ 0 \\ 0 \\ 0 \end{bmatrix}$$

$$= \left(\frac{k_1}{m_1}\right)\left(\frac{bm_1 s + k_2 m_1}{m_1 m_2 s^4 + b(m_1+m_2)s^3 + (k_1 m_2 + k_2 m_1 + k_2 m_2)s^2 + bk_1 s + k_1 k_2}\right) \tag{2.110}$$

Similarly, applying (2.109) to the alternate state-space model (2.96) gives

$$\frac{Y(s)}{U(s)} = [0\ 0\ 0\ 1]\left(s\begin{bmatrix}1&0&0&0\\0&1&0&0\\0&0&1&0\\0&0&0&1\end{bmatrix} - \begin{bmatrix}-\left(\frac{b}{m_1}+\frac{b}{m_2}\right) & -\left(\frac{k_1+k_2}{m_1}+\frac{k_2}{m_2}\right) & 0 & \frac{k_1}{m_1}\\ 1 & 0 & 0 & 0\\ -\frac{b}{m_2} & -\frac{k_2}{m_2} & 0 & 0\\ 0 & 0 & 1 & 0\end{bmatrix}\right)^{-1}\begin{bmatrix}-\frac{k_1}{m_1}\\0\\0\\0\end{bmatrix}$$

$$= [0\ 0\ 0\ 1]\begin{bmatrix}s+\left(\frac{b}{m_1}+\frac{b}{m_2}\right) & \left(\frac{k_1+k_2}{m_1}+\frac{k_2}{m_2}\right) & 0 & -\frac{k_1}{m_1}\\ -1 & s & 0 & 0\\ \frac{b}{m_2} & \frac{k_2}{m_2} & s & 0\\ 0 & 0 & -1 & s\end{bmatrix}^{-1}\begin{bmatrix}-\frac{k_1}{m_1}\\0\\0\\0\end{bmatrix}$$

$$= \left(-\frac{k_1}{m_1}\right)\left(-\frac{bm_1s+k_2m_1}{m_1m_2s^4+b(m_1+m_2)s^3+(k_1m_2+k_2m_1+k_2m_2)s^2+bk_1s+k_1k_2}\right) \tag{2.111}$$

That the final results in (2.110) and (2.111) are equal points to an important fact. The transfer function for a given system is unique even though the state-space model is not.

Note that the lower case version of the b, c, and d matrices are used in (2.109). The reason for this is that by definition, the transfer function relates one input to one output. If there are multiple inputs or multiple outputs to deal with, then a matrix form of the transfer function, denoted by H(s), is available.

$$H(s) = C(sI - A)^{-1}B + D \tag{2.112}$$

where

$$H(s) = \begin{bmatrix}H_{11}(s) & H_{12}(s) & \cdots & H_{1p}(s)\\ H_{21}(s) & H_{22}(s) & \cdots & H_{2p}(s)\\ \vdots & \vdots & \ddots & \vdots\\ H_{q1}(s) & H_{q2} & \cdots & H_{qp}(s)\end{bmatrix} \tag{2.113}$$

and $H_{ij}(s)$ is the transfer function from the jth input to the ith output.

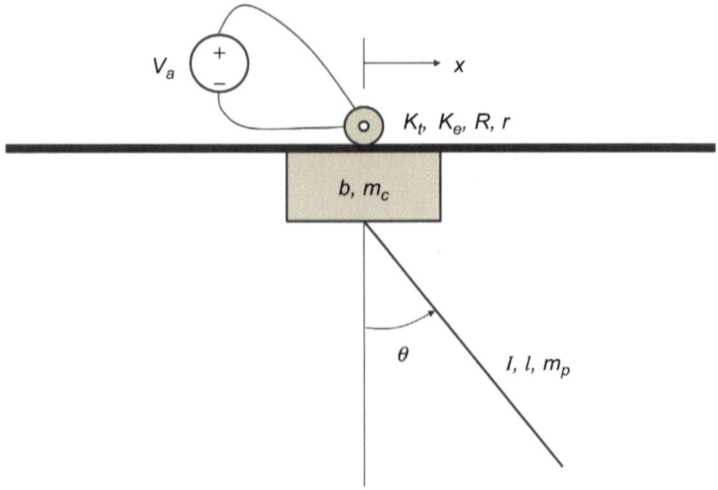

Figure 2.46 Model for the hanging crane.

2.4.3.1 MATLAB Example: Converting a State-Space Model to a Transfer Function for a Hanging Crane

The MATLAB command for converting a state-space model to transfer function form is `ss2tf`. In this example, the model of the hanging crane[8] is used, as shown in Figure 2.46. In this system, the input is the voltage V_a applied to the motor, whose rotational motion is converted to linear motion of the cart, which in turn causes the pendulum to swing. The cart position is denoted by x and the pendulum angle by θ. The constants for the system are K_t, the torque constant; K_e, the electric constant; R, the coil resistance; r, the radius of the wheel; b, the friction of the cart; m_c, the mass of the cart; I, the inertia of the pendulum; l, the length of the pendulum; and m_p, the mass of the pendulum. It is assumed the inductance of the motor coil, motor inertia, motor friction, and pendulum friction are small enough to be ignored.

The dynamic equations for the system are

$$
\begin{gathered}
(I + m_p l^2)\ddot{\theta} + m_p g l \sin\theta = -m_p l \ddot{x} \cos\theta \\
(m_c + m_p)\ddot{x} + b\dot{x} + m_p l \ddot{\theta} \cos\theta - m_p l \dot{\theta}^2 \sin\theta = \frac{K_t}{rR}\left(V_a - \frac{K_e}{r}\dot{x}\right)
\end{gathered}
\tag{2.114}
$$

[8] The model here is adopted from Franklin, Powell, and Emami-Naeini (2010).

Converting this model to state-space form and defining the A, B, C, and D matrices is impossible because the system is nonlinear. The A matrix multiplies the x vector, and there is no means of taking a sine or cosine of a state in that scenario as is needed in this system. Fortunately, we can linearize (2.114) so that the matrices for state-space representation can be obtained. (Linearization and nonlinear modeling are discussed in Chapter 4.) Linearizing about $\theta = 0$ and $\dot{\theta} = 0$ yields

$$\begin{aligned}(I + m_p l^2)\ddot{\theta} + m_p g l \theta &= -m_p l \ddot{x} \\ (m_c + m_p)\ddot{x} + b\dot{x} + m_p l \ddot{\theta} &= \frac{K_t}{rR}\left(V_a - \frac{K_e}{r}\dot{x}\right)\end{aligned} \tag{2.115}$$

Solving the first equation in (2.115) for $\ddot{\theta}$, plugging the result into the second equation, and solving for $\ddot{x}$ gives

$$\begin{aligned}\ddot{x} = {} & \frac{\frac{K_t}{rR}(I + m_p l^2)}{I(m_c + m_p) + m_c m_p l^2} V_a - \frac{\left(\frac{K_t K_e}{r^2 R} + b\right)(I + m_p l^2)}{I(m_c + m_p) + m_c m_p l^2} \dot{x} \\ & - \frac{m_p^2 l^2}{I(m_c + m_p) + m_c m_p l^2} \theta\end{aligned} \tag{2.116}$$

Similarly, solving the first equation in (2.115) for $\ddot{x}$, plugging the results into the second equation, and solving for $\ddot{\theta}$ gives

$$\begin{aligned}\ddot{\theta} = {} & -\frac{\frac{K_t}{rR} m_p l}{(m_c + m_p)I + m_c m_p l^2} V_a - \frac{(m_c + m_p) m_p g l}{(m_c + m_p)I + m_c m_p l^2} \theta \\ & + \frac{\left(\frac{K_t K_e}{r^2 R} + b\right)(m_p l)}{(m_c + m_p)I + m_c m_p l^2} \dot{x}\end{aligned} \tag{2.117}$$

Choosing the states to be

$$\begin{aligned}x_1 &= x \\ x_2 &= \dot{x} \\ x_3 &= \theta \\ x_4 &= \dot{\theta}\end{aligned} \tag{2.118}$$

and the outputs to be

$$\begin{aligned}y_1 &= x \\ y_2 &= \theta\end{aligned} \tag{2.119}$$

gives the state-space model

$$
\begin{bmatrix} \dot{x}_1 \\ \dot{x}_2 \\ \dot{x}_3 \\ \dot{x}_4 \end{bmatrix} =
\begin{bmatrix}
0 & 1 & 0 & 0 \\
0 & -\dfrac{\left(\dfrac{K_t K_e}{r^2 R} + b\right)(I + m_p l^2)}{I(m_c + m_p) + m_c m_p l^2} & -\dfrac{m_p^2 l^2}{I(m_c + m_p) + m_c m_p l^2} & 0 \\
0 & 0 & 0 & 1 \\
0 & -\dfrac{\left(\dfrac{K_t K_e}{r^2 R} + b\right)(I + m_p l^2)}{I(m_c + m_p) + m_c m_p l^2} & -\dfrac{(m_c + m_p) m_p g l}{(m_c + m_p) I + m_c m_p l^2} & 0
\end{bmatrix}
\begin{bmatrix} x_1 \\ x_2 \\ x_3 \\ x_4 \end{bmatrix}
+ \begin{bmatrix}
0 \\
\dfrac{\dfrac{K_t}{rR}(I + m_p l^2)}{I(m_c + m_p) + m_c m_p l^2} \\
0 \\
-\dfrac{\dfrac{K_t}{rR} m_p l}{(m_c + m_p) I + m_c m_p l^2}
\end{bmatrix} V_a
$$

$$
\begin{bmatrix} y_1 \\ y_2 \end{bmatrix} = \begin{bmatrix} 1 & 0 & 0 & 0 \\ 0 & 0 & 1 & 0 \end{bmatrix} \begin{bmatrix} x_1 \\ x_2 \\ x_3 \\ x_4 \end{bmatrix} + \begin{bmatrix} 0 \\ 0 \end{bmatrix} V_a \tag{2.120}
$$

The MATLAB code below generates the state-space model and converts it to transfer function form.

```
% hanging_crane.m

close all
clear all

Kt = 5.9e-3;        % Nm/A
Ke = 5.9e-3;
r = 0.02;           % m
```

```
R = 10;              % Ohms
b = 0.1;             % Ns/m
I = 0.005;           % kg m^2
mc = 0.5;            % kg
mp = 0.03;           % kg
l = 0.1;             % m
g = 9.8;             % m/s^2

d1 = I*(mc+mp)+mc*mp*l^2;
A=[0,1,0,0;0,-(Kt*Ke/(r^2*R)+b)*(I+mp*l^2)/d1,-mp^2*l^2/d1,0;0,0,
0,1;0,-(Kt*Ke/(r^2*R)+b)*(I+mp*l^2)/d1,-(mc+mp)*mp*g*l/d1,0];
B = [0; Kt/(r*R)*(I+mp*l^2)/d1; 0; -Kt/(r*R)*mp*l/d1];
C = [1, 0, 0, 0; 0, 0, 1, 0];
D = [0; 0];

[num,den] = ss2tf(A,B,C,D)

tf1 = tf(num(1,:),den)
tf2 = tf(num(2,:),den)

subplot(2,1,1), step(tf1,60), ylabel('Cart Position (m)')
subplot(2,1,2), step(tf2,60), ylabel('Pendulum Angle (rad)')
```

In the above code, after initialization of all the constants, the `A`, `B`, `C`, and `D` matrices are defined. The variable `d1` is used to hold the denominator of several matrix entries, simplifying the code.

Next, the `ss2tf` command is used to obtain the transfer function. There are a few items to note about this command. First, it takes the A, B, C, and D matrices as its arguments rather than the state-space model as generated by the `ss` command. Similarly, it returns the numerator and denominator coefficients of the transfer function rather than the transfer function model as generated by the `tf` command. Finally, the numerator vector returned is of appropriate size, meaning the numerator has as many rows as outputs. The result of this command is

```
num =
           0         0    0.0558   -0.0000    0.3108
           0         0   -0.0316   -0.0180    0.0000
den =
      1.0000    0.2058    5.5650    1.1444         0
```

In the next two lines, the `tf` command is used to generate two separate transfer functions: one relating the cart position (state x_1) to the voltage applied and the other relating the pendulum angle (state x_3) to the voltage applied. The two transfer functions returned are

```
tf1 =

    0.05584 s^2 - 3.72e-17 s + 0.3108
  ----------------------------------------
  s^4 + 0.2058 s^3 + 5.565 s^2 + 1.144 s

Continuous-time transfer function.

tf2 =

    -0.03161 s^2 - 0.01799 s + 1.049e-17
  ----------------------------------------
  s^4 + 0.2058 s^3 + 5.565 s^2 + 1.144 s

Continuous-time transfer function.
```

The final two lines of code plot and label the step response of the system for 60 seconds. The resulting plot is shown in Figure 2.47.

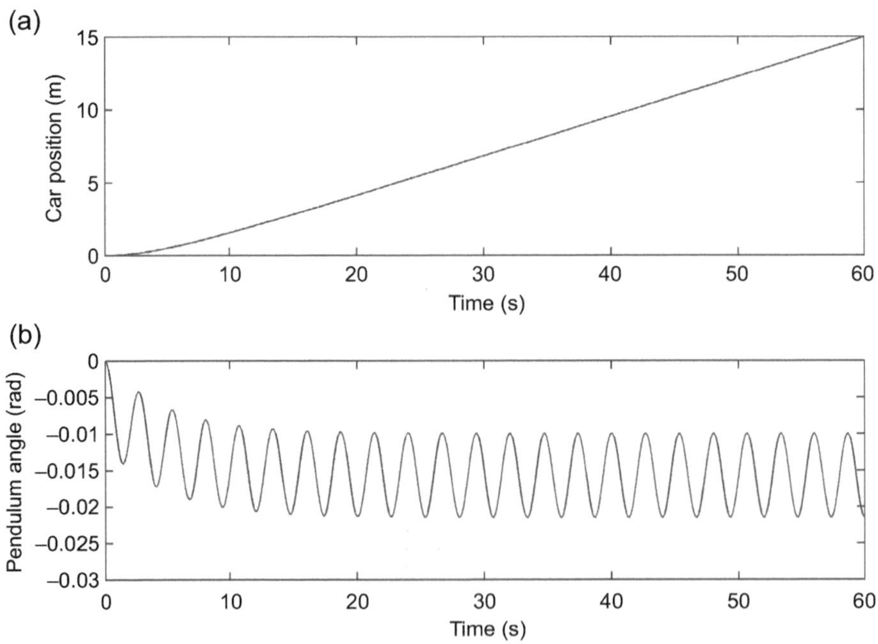

Figure 2.47 The response of the hanging crane to a 1 V input signal.

As shown in the figure, the cart position increases linearly (after an initial transient) and will continue increasing as long as the voltage is applied. The pendulum angle becomes negative and starts to oscillate around an offset. Because we assumed a frictionless pendulum, this oscillation will continue, and the average angle will remain at a constant, as we expect the pendulum to behave when the cart is moving at a constant speed.

To obtain a state-space model from a transfer function, knowing that state-space representation is not unique (there are infinitely many ways to represent a system), we have options for which approach to take. These options are discussed in the next section.

There is one final note to make before concluding this section. We have been focusing on continuous-time systems in the examples and MATLAB code. Discrete-time systems are treated in the same manner, except z is used in place of s and $x[n+1]$ is used in place of $\dot{x}$.

2.4.4 Canonical Forms

As stated earlier, there are infinitely many ways to represent a system in state-space form. However, not all of them are equally useful. A few state-space representations are used more often because they reveal certain features of the system or put it in a form that makes it easier to perform certain tasks, such as designing controllers for the system. These particular representations are called **canonical forms**. In this section, we will discuss a number of these forms: controllable, observable, phase variable, modal (or diagonal), and Jordan. Unless otherwise noted, in discussing these forms, we start with the SISO system equations

$$\begin{aligned}\frac{d^N y}{dt^N} + a_1\frac{d^{N-1}y}{dt^{N-1}} + \cdots + a_{N-1}\frac{dy}{dt} + a_N y = {} & b_0\frac{d^N u}{dt^N} + b_1\frac{d^{N-1}u}{dt^{N-1}} + \cdots \\ & + b_{N-1}\frac{du}{dt} + b_N u\end{aligned} \tag{2.121}$$

and the resulting transfer function

$$\frac{Y(s)}{U(s)} = \frac{b_0 s^N + b_1 s^{N-1} + \cdots + b_{N-1}s + b_N}{s^N + a_1 s^{N-1} + \cdots + a_{N-1}s + a_N} \tag{2.122}$$

It is assumed that the system is proper, that is, the highest power of s in the numerator is not greater than the highest power of s in the denominator. This is a characteristic of real-world, physical systems.

2.4.4.1 Controllable Canonical Form

Controllable canonical form is useful in analyzing and designing control systems because this form guarantees controllability. A system is controllable if it can move from any state to any other state in finite time. A controllable system can be made to follow any trajectory the designer desires.

$$\begin{bmatrix} \dot{x}_1 \\ \dot{x}_2 \\ \vdots \\ \dot{x}_{N-1} \\ \dot{x}_N \end{bmatrix} = \begin{bmatrix} 0 & 1 & 0 & \cdots & 0 \\ 0 & 0 & 1 & \cdots & 0 \\ \vdots & \vdots & \vdots & \ddots & \vdots \\ 0 & 0 & 0 & \cdots & 1 \\ -a_N & -a_{N-1} & -a_{N-2} & \cdots & -a_1 \end{bmatrix} \begin{bmatrix} x_1 \\ x_2 \\ \vdots \\ x_{N-1} \\ x_N \end{bmatrix} + \begin{bmatrix} 0 \\ 0 \\ \vdots \\ 0 \\ 1 \end{bmatrix} u$$

$$y = \begin{bmatrix} b_N - a_N b_0 & b_{N-1} - a_{N-1} b_0 & \cdots & b_2 - a_2 b_0 & b_1 - a_1 b_0 \end{bmatrix} \begin{bmatrix} x_1 \\ x_2 \\ \vdots \\ x_{N-1} \\ x_N \end{bmatrix} + [b_0]u \tag{2.123}$$

To see where this form comes from, let $X(s)$ be defined as

$$X(s) = \frac{1}{s^N + a_1 s^{N-1} + \cdots + a_{N-1}s + a_N} U(s) \tag{2.124}$$

Then

$$Y(s) = (b_0 s^N + b_1 s^{N-1} + \cdots + b_{N-1}s + b_N)X(s) \tag{2.125}$$

Using the differentiation property of the Laplace transform, converting (2.124) and (2.125) into the time domain gives

$$\frac{d^N x}{dt^N} + a_1 \frac{d^{N-1} x}{dt^{N-1}} + \cdots + a_{N-1} \frac{dx}{dt} + a_N x = u \tag{2.126}$$

and

$$y = b_0 \frac{d^N x}{dt^N} + b_1 \frac{d^{N-1} x}{dt^{N-1}} + \cdots + b_{N-1} \frac{dx}{dt} + b_N x \tag{2.127}$$

Finally, define the states to be x and its derivatives

$$x_i = \frac{d^{i-1} x}{dt^{i-1}} \tag{2.128}$$

where $i = 1, \ldots, N$. Solving for $\dot{x}_N$ using (2.126) gives the state equation in (2.123). The output equation is obtained using (2.127) and substituting $\dot{x}_N$ for $\frac{d^N x}{dt^N}$ to get

$$\begin{aligned} y = {} & b_0(-a_N x_1 - a_{N-1} x_2 - \cdots - a_2 x_{N-1} - a_1 x_N + u) + b_1 x_N \\ & + b_2 x_{N-1} + \cdots + b_{N-1} x_2 + b_N x_1 \end{aligned} \tag{2.129}$$

Rearranging (2.129) gives the output equation in (2.123).

2.4.4.2 Observable Canonical Form

Observable canonical form is also useful in analyzing and designing control systems because this form guarantees observability. A system is observable if all its states can be determined by the output. Observability is useful because it means the initial condition of a system can be back calculated from what can be physically measured.

$$\begin{bmatrix} \dot{x}_1 \\ \dot{x}_2 \\ \vdots \\ \dot{x}_{N-1} \\ \dot{x}_N \end{bmatrix} = \begin{bmatrix} 0 & 0 & \cdots & 0 & -a_N \\ 1 & 0 & \cdots & 0 & -a_{N-1} \\ \vdots & \vdots & \ddots & \vdots & \vdots \\ 0 & 0 & \cdots & 0 & -a_2 \\ 0 & 0 & \cdots & 1 & -a_1 \end{bmatrix} \begin{bmatrix} x_1 \\ x_2 \\ \vdots \\ x_{N-1} \\ x_N \end{bmatrix} + \begin{bmatrix} b_N - a_N b_0 \\ b_{N-1} - a_{N-1} b_0 \\ \vdots \\ b_2 - a_2 b_0 \\ b_1 - a_1 b_0 \end{bmatrix} u$$

$$y = \begin{bmatrix} 0 & 0 & \cdots & 0 & 1 \end{bmatrix} \begin{bmatrix} x_1 \\ x_2 \\ \vdots \\ x_{N-1} \\ x_N \end{bmatrix} + [b_0] u \tag{2.130}$$

Notice that observable canonical form can be obtained from the controllable canonical form simply by transposing the A matrix and by transposing and swapping the b and c vectors.

2.4.4.3 Phase Variable Canonical Form

$$\begin{bmatrix} \dot{x}_1 \\ \dot{x}_2 \\ \vdots \\ \dot{x}_{N-1} \\ \dot{x}_N \end{bmatrix} = \begin{bmatrix} 0 & 1 & 0 & \cdots & 0 \\ 0 & 0 & 1 & \cdots & 0 \\ \vdots & \vdots & \vdots & \ddots & \vdots \\ 0 & 0 & 0 & \cdots & 1 \\ -a_N & -a_{N-1} & -a_{N-2} & \cdots & -a_1 \end{bmatrix} \begin{bmatrix} x_1 \\ x_2 \\ \vdots \\ x_{N-1} \\ x_N \end{bmatrix} + \begin{bmatrix} 0 \\ 0 \\ \vdots \\ 0 \\ k \end{bmatrix} u$$

$$y = \begin{bmatrix} 1 & 0 & \cdots & 0 & 0 \end{bmatrix} \begin{bmatrix} x_1 \\ x_2 \\ \vdots \\ x_{N-1} \\ x_N \end{bmatrix} + [0]u \tag{2.131}$$

Phase variable canonical form is a special case of controllable canonical form and assumes the system has form

$$\frac{d^N y}{dt^N} + a_1 \frac{d^{N-1} y}{dt^{N-1}} + \cdots + a_{N-1} \frac{dy}{dt} + a_N y = ku \tag{2.132}$$

and corresponding transfer function

$$\frac{Y(s)}{U(s)} = \frac{k}{s^N + a_1 s^{N-1} + \cdots + a_{N-1}s + a_N} \tag{2.133}$$

2.4.4.4 Modal (or Diagonal) Canonical Form

Modal canonical form allows one to immediately identify the poles of the system (i.e., the roots of the denominator). Poles of the system play a large part in determining the stability of a system as discussed in Section 3.4. Modal canonical form assumes the poles are distinct. Jordan canonical form (discussed in the following section) allows for repeated poles.

$$\begin{bmatrix} \dot{x}_1 \\ \dot{x}_2 \\ \vdots \\ \dot{x}_{N-1} \\ \dot{x}_N \end{bmatrix} = \begin{bmatrix} -p_1 & 0 & \cdots & 0 & 0 \\ 0 & -p_2 & \cdots & 0 & 0 \\ \vdots & \vdots & \ddots & \vdots & \vdots \\ 0 & 0 & \cdots & -p_{N-1} & 0 \\ 0 & 0 & \cdots & 0 & -p_N \end{bmatrix} \begin{bmatrix} x_1 \\ x_2 \\ \vdots \\ x_{N-1} \\ x_N \end{bmatrix} + \begin{bmatrix} 1 \\ 1 \\ \vdots \\ 1 \\ 1 \end{bmatrix} u$$

$$y = \begin{bmatrix} c_1 & c_2 & \cdots & c_{N-1} & c_N \end{bmatrix} \begin{bmatrix} x_1 \\ x_2 \\ \vdots \\ x_{N-1} \\ x_N \end{bmatrix} + [b_0]u \tag{2.134}$$

Modal canonical form assumes that the transfer function can be written as

$$\frac{Y(s)}{U(s)} = b_0 + \frac{c_1}{s + p_1} + \frac{c_2}{s + p_2} + \cdots + \frac{c_{N-1}}{s + p_{N-1}} + \frac{c_N}{s + p_N} \tag{2.135}$$

To obtain the canonical form, write the output as

$$\begin{aligned} Y(s) = b_0 U(s) &+ \frac{c_1}{s + p_1} U(s) + \frac{c_2}{s + p_2} U(s) + \cdots + \frac{c_{N-1}}{s + p_{N-1}} U(s) \\ &+ \frac{c_N}{s + p_N} U(s) \end{aligned} \tag{2.136}$$

and define the states to be

$$X_i(s) = \frac{1}{s + p_i} U(s) \tag{2.137}$$

for $i = 1, \ldots, N$. In the time domain, the expression becomes

$$\frac{dx_i}{dt} + p_i x_i = u \tag{2.138}$$

Thus, the result is the state-space representation in (2.134).

2.4.4.5 Jordan Canonical Form

As with modal canonical form, Jordan canonical form allows one to easily identify the poles of the system because they are arranged along the diagonal of the A matrix. The difference between them is that this form accommodates repeated poles.

$$\begin{bmatrix} \dot{x}_1 \\ \dot{x}_2 \\ \vdots \\ \dot{x}_{r-1} \\ \dot{x}_r \\ \dot{x}_{r+1} \\ \dot{x}_{r+2} \\ \vdots \\ \dot{x}_{N-1} \\ \dot{x}_N \end{bmatrix} = \left[\begin{array}{ccccc|ccccc} -p_1 & 1 & \cdots & 0 & 0 & 0 & 0 & \cdots & 0 & 0 \\ 0 & -p_1 & \cdots & 0 & 0 & 0 & 0 & \cdots & 0 & 0 \\ \vdots & \vdots & \ddots & \vdots & \vdots & \vdots & \vdots & \ddots & \vdots & \vdots \\ 0 & 0 & \cdots & -p_1 & 1 & 0 & 0 & \cdots & 0 & 0 \\ 0 & 0 & \cdots & 0 & -p_1 & 0 & 0 & \cdots & 0 & 0 \\ \hline 0 & 0 & \cdots & 0 & 0 & -p_{r+1} & 0 & \cdots & 0 & 0 \\ 0 & 0 & \cdots & 0 & 0 & 0 & -p_{r+2} & \cdots & 0 & 0 \\ \vdots & \vdots & \ddots & \vdots & \vdots & \vdots & \vdots & \ddots & \vdots & \vdots \\ 0 & 0 & \cdots & 0 & 0 & 0 & 0 & \cdots & -p_{N-1} & 0 \\ 0 & 0 & \cdots & 0 & 0 & 0 & 0 & \cdots & 0 & -p_N \end{array}\right] \begin{bmatrix} x_1 \\ x_2 \\ \vdots \\ x_{r-1} \\ x_r \\ x_{r+1} \\ x_{r+2} \\ \vdots \\ x_{N-1} \\ x_N \end{bmatrix} + \begin{bmatrix} 0 \\ 0 \\ \vdots \\ 0 \\ 1 \\ 1 \\ 1 \\ \vdots \\ 1 \\ 1 \end{bmatrix} u$$

$$y = \begin{bmatrix} c_{1r} & c_{1r-1} & \cdots & c_{12} & c_{11} & c_{r+1} & c_{r+2} & \cdots & c_{N-1} & c_N \end{bmatrix} \begin{bmatrix} x_1 \\ x_2 \\ \vdots \\ x_{r-1} \\ x_r \\ x_{r+1} \\ x_{r+2} \\ \vdots \\ x_{N-1} \\ x_N \end{bmatrix} + [b_0]u \tag{2.139}$$

Jordan canonical form assumes the transfer function can be rearranged as

$$\begin{aligned} \frac{Y(s)}{U(s)} &= \frac{b_0 s^N + b_1 s^{N-1} + \cdots + b_{N-1}s + b_N}{(s+p_1)^r (s+p_{r+1}) \cdots (s+p_N)} \\ &= b_0 + \frac{c_{11}}{s+p_1} + \frac{c_{12}}{(s+p_1)^2} + \frac{c_{1r}}{(s+p_1)^r} + \frac{c_{r+1}}{s+p_{r+1}} + \cdots + \frac{c_N}{s+p_N} \end{aligned} \tag{2.140}$$

2.4.4.6 Chained Form

$$\begin{bmatrix} \dot{x}_1 \\ \dot{x}_2 \\ \dot{x}_3 \\ \vdots \\ \dot{x}_{N-1} \\ \dot{x}_N \end{bmatrix} = \begin{bmatrix} 1 \\ 0 \\ x_2 \\ \vdots \\ x_{N-2} \\ x_{N-1} \end{bmatrix} u_1 + \begin{bmatrix} 0 \\ 1 \\ 0 \\ \vdots \\ 0 \\ 0 \end{bmatrix} u_2 \tag{2.141}$$

This form is known as (2,N) chained form because it has two inputs and N states. It differs from the previous canonical forms in that it does not use A, B, C, and D matrices and can be used to model nonlinear systems.

2.4.4.7 Application Examples

Let us now look at the various canonical forms in the context of system examples. First we return to the transfer function of the car suspension system, rearranged in the form of (2.122).

$$\frac{Y(s)}{U(s)} = \frac{\frac{bk_1}{m_1 m_2} s + \frac{k_1 k_2}{m_1 m_2}}{s^4 + \left(\frac{b}{m_1} + \frac{b}{m_2}\right) s^3 + \left(\frac{k_1+k_2}{m_1} + \frac{k_2}{m_2}\right) s^2 + \frac{bk_1}{m_1 m_2} s + \frac{k_1 k_2}{m_1 m_2}} \tag{2.142}$$

The controllable canonical form of the state-space model is

$$\begin{bmatrix} \dot{x}_1 \\ \dot{x}_2 \\ \dot{x}_3 \\ \dot{x}_4 \end{bmatrix} = \begin{bmatrix} 0 & 1 & 0 & 0 \\ 0 & 0 & 1 & 0 \\ 0 & 0 & 0 & 1 \\ -\frac{k_1 k_2}{m_1 m_2} & -\frac{bk_1}{m_1 m_2} & -\left(\frac{k_1 + k_2}{m_1} + \frac{k_2}{m_2}\right) & -\left(\frac{b}{m_1} + \frac{b}{m_2}\right) \end{bmatrix} \begin{bmatrix} x_1 \\ x_2 \\ x_3 \\ x_4 \end{bmatrix} + \begin{bmatrix} 0 \\ 0 \\ 0 \\ 1 \end{bmatrix} u$$

$$y = \begin{bmatrix} \frac{k_1 k_2}{m_1 m_2} & \frac{bk_1}{m_1 m_2} & 0 & 0 \end{bmatrix} \begin{bmatrix} x_1 \\ x_2 \\ x_3 \\ x_4 \end{bmatrix} + [0]u \tag{2.143}$$

The observable canonical form of the state-space model is

$$\begin{bmatrix} \dot{x}_1 \\ \dot{x}_2 \\ \dot{x}_3 \\ \dot{x}_4 \end{bmatrix} = \begin{bmatrix} 0 & 0 & 0 & -\dfrac{k_1 k_2}{m_1 m_2} \\ 1 & 0 & 0 & -\dfrac{bk_1}{m_1 m_2} \\ 0 & 1 & 0 & -\left(\dfrac{k_1 + k_2}{m_1} + \dfrac{k_2}{m_2}\right) \\ 0 & 0 & 1 & -\left(\dfrac{b}{m_1} + \dfrac{b}{m_2}\right) \end{bmatrix} \begin{bmatrix} x_1 \\ x_2 \\ x_3 \\ x_4 \end{bmatrix} + \begin{bmatrix} \dfrac{k_1 k_2}{m_1 m_2} \\ \dfrac{bk_1}{m_1 m_2} \\ 0 \\ 0 \end{bmatrix} u \tag{2.144}$$

$$y = \begin{bmatrix} 0 & 0 & 0 & 1 \end{bmatrix} \begin{bmatrix} x_1 \\ x_2 \\ x_3 \\ x_4 \end{bmatrix} + [0]u$$

Now let us consider a second-order system and return to the pendulum model introduced earlier in the chapter (see Figure 2.19). However, here we will include a torque input u around the axis of rotation in the same direction as θ. With this added input, the system equation becomes

$$u - mgl \sin \theta - b\dot{\theta} = ml^2\ddot{\theta} \tag{2.145}$$

Using $\sin \theta \approx \theta$ for small values of θ, the linear equation is

$$u - mgl\theta - b\dot{\theta} = ml^2\ddot{\theta} \tag{2.146}$$

The transfer function of this system taking θ to be the output gives

$$\frac{\Theta(s)}{U(s)} = \frac{\frac{1}{ml^2}}{s^2 + \frac{b}{ml^2}s + \frac{g}{l}} \tag{2.147}$$

The controllable canonical form of the pendulum state-space model is

$$\begin{bmatrix} \dot{x}_1 \\ \dot{x}_2 \end{bmatrix} = \begin{bmatrix} 0 & 1 \\ -\frac{g}{l} & -\frac{b}{ml^2} \end{bmatrix} \begin{bmatrix} x_1 \\ x_2 \end{bmatrix} + \begin{bmatrix} 0 \\ 1 \end{bmatrix} u$$

$$y = \begin{bmatrix} \frac{1}{ml^2} & 0 \end{bmatrix} \begin{bmatrix} x_1 \\ x_2 \end{bmatrix} + [0]u \tag{2.148}$$

What do the states x_1 and x_2 represent? In the case of the car suspension, the states had no real-world meaning. They were simply a string of integrated variables. In this case, however, they are related to the angle of the pendulum by a simple relationship. We know the output of the system is the angle θ. Combining this fact with the output equation of (2.148) gives

$$y = \frac{1}{ml^2}x_1 = \theta \tag{2.149}$$

Therefore,

$$\begin{aligned} x_1 &= ml^2\theta \\ x_2 &= ml^2\dot{\theta} \end{aligned} \tag{2.150}$$

are simply scaled versions of the angle and its derivative.

Observable canonical form is obtained by transposing the A matrix and transposing and swapping the b and c vectors.

$$\begin{bmatrix} \dot{x}_1 \\ \dot{x}_2 \end{bmatrix} = \begin{bmatrix} 0 & -\frac{g}{l} \\ 1 & -\frac{b}{ml^2} \end{bmatrix} \begin{bmatrix} x_1 \\ x_2 \end{bmatrix} + \begin{bmatrix} \frac{1}{ml^2} \\ 0 \end{bmatrix} u$$

$$y = \begin{bmatrix} 0 & 1 \end{bmatrix} \begin{bmatrix} x_1 \\ x_2 \end{bmatrix} + [0]u \tag{2.151}$$

In this case, the states are

$$\begin{aligned} x_2 &= \theta \\ x_1 &= \dot{\theta} + \frac{b}{ml^2}\theta \end{aligned} \tag{2.152}$$

The diagonal canonical form can be obtained after first applying partial fraction expansion to the transfer function.

$$\frac{\Theta(s)}{U(s)} = \frac{\frac{1}{ml^2\sqrt{\left(\frac{b}{ml^2}\right)^2 - 4\frac{g}{l}}}}{s + \left(\frac{b}{2ml^2} - \sqrt{\left(\frac{b}{2ml^2}\right)^2 - \frac{g}{l}}\right)} + \frac{-\frac{1}{ml^2\sqrt{\left(\frac{b}{ml^2}\right)^2 - 4\frac{g}{l}}}}{s + \left(\frac{b}{2ml^2} + \sqrt{\left(\frac{b}{2ml^2}\right)^2 - \frac{g}{l}}\right)} \tag{2.153}$$

Then the state-space model becomes

$$\begin{bmatrix} \dot{x}_1 \\ \dot{x}_2 \end{bmatrix} = \begin{bmatrix} -\left(\frac{b}{2ml^2} - \sqrt{\left(\frac{b}{2ml^2}\right)^2 - \frac{g}{l}}\right) & 0 \\ 0 & -\left(\frac{b}{2ml^2} + \sqrt{\left(\frac{b}{2ml^2}\right)^2 - \frac{g}{l}}\right) \end{bmatrix} \begin{bmatrix} x_1 \\ x_2 \end{bmatrix} + \begin{bmatrix} 1 \\ 1 \end{bmatrix} u$$

$$y = \begin{bmatrix} \frac{1}{ml^2\sqrt{\left(\frac{b}{ml^2}\right)^2 - 4\frac{g}{l}}} & -\frac{1}{ml^2\sqrt{\left(\frac{b}{ml^2}\right)^2 - 4\frac{g}{l}}} \end{bmatrix} \begin{bmatrix} x_1 \\ x_2 \end{bmatrix} + [0]u \tag{2.154}$$

Note that (2.154) is the diagonal form of the state-space model and assumes that the transfer function denominator has distinct roots. In other words, the system is overdamped or underdamped but not critically damped, meaning

$$\left(\frac{b}{ml^2}\right)^2 - 4\frac{g}{l} \neq 0 \tag{2.155}$$

Also note that in the underdamped case, where the denominator has complex roots, the elements of the matrices are complex. We will discuss the meaning of complex matrices in the next section, which covers eigenvalues and eigenvectors.

For the critically damped system, where equality holds in (2.155), the state-space model can be expressed in Jordan canonical form. The transfer function becomes

$$\frac{\Theta(s)}{U(s)} = \frac{\frac{1}{ml^2}}{\left(s + \frac{b}{2ml^2}\right)^2} \tag{2.156}$$

Then the Jordan canonical form of the state-space model is

$$\begin{bmatrix} \dot{x}_1 \\ \dot{x}_2 \end{bmatrix} = \begin{bmatrix} -\dfrac{b}{2ml^2} & 1 \\ 0 & -\dfrac{b}{2ml^2} \end{bmatrix} \begin{bmatrix} x_1 \\ x_2 \end{bmatrix} + \begin{bmatrix} 0 \\ 1 \end{bmatrix} u$$
$$y = \begin{bmatrix} \dfrac{1}{ml^2} & 0 \end{bmatrix} \begin{bmatrix} x_1 \\ x_2 \end{bmatrix} + [0]u \tag{2.157}$$

The states are related to the pendulum angle by

$$\begin{aligned} x_1 &= ml^2\theta \\ x_2 &= ml^2\dot{\theta} + \frac{b}{2}\theta \end{aligned} \tag{2.158}$$

The modal and Jordan canonical forms are directly related to the eigenvalues of the system; we will discuss this connection in the next section.

2.4.5 Eigenvalues and Eigenvectors

Eigenvalues and eigenvectors are a familiar topic to anyone who has studied linear algebra and matrices. Recall the definition

$$A\hat{x} = \lambda\hat{x} \tag{2.159}$$

where A is an $N \times N$ matrix, $\hat{x}$ is an $N \times 1$ vector, and λ is a scalar. In (2.159), λ is an **eigenvalue** of A, and $\hat{x}$ is an **eigenvector** of A. The eigenvalues of A are found by solving the **characteristic equation** of A defined as

$$\det(\lambda I - A) = 0 \tag{2.160}$$

where I is the $N \times N$ identity matrix. The eigenvector associated with an eigenvalue is found by solving the equation

$$(A - \lambda I)\widehat{x} = 0 \tag{2.161}$$

which is simply a rearrangement of (2.159).

The procedure for finding eigenvalues and eigenvectors is often taught and drilled during a course in linear algebra, and therefore the process becomes ingrained. However, what do these concepts mean in the context of dynamical systems? This question is the topic of this section.

The first thing to note is that the use of the same matrix notation A in (2.159) and in the state-space models is not a coincidence. The expression in (2.159) can be interpreted as describing the evolution of the system when there is no input ($u = 0$). In this view,

$$\begin{aligned}\dot{x} &= A\widehat{x} \\ &= \lambda\widehat{x}\end{aligned} \tag{2.162}$$

means that the direction in which the system is moving ($\dot{x}$), is in the same direction it is already pointed.

Let's look at these concepts in the context of an example and revisit phase plots to aid in understanding. Recall the linearized model for the pendulum with no input torque.

$$-mgl\theta - b\dot{\theta} = ml^2\ddot{\theta} \tag{2.163}$$

The resulting state-space representation is

$$\begin{bmatrix} \dot{\theta} \\ \ddot{\theta} \end{bmatrix} = \begin{bmatrix} 0 & 1 \\ -\dfrac{g}{l} & -\dfrac{b}{ml^2} \end{bmatrix} \begin{bmatrix} \theta \\ \dot{\theta} \end{bmatrix} \tag{2.164}$$

Note that we are only investigating the state of the system in the absence of inputs. Thus, matrix $b = 0$, and we do not need to consider the c and d matrices in the output equation of the state-space model.

Now consider three cases: **overdamped**, **critically damped**, and **underdamped** corresponding to distinct real, repeated real, and complex conjugate solutions of the characteristic equation. For all three cases, we will use MATLAB to help visualize the behavior of the system.

2.4.5.1 MATLAB Example: Eigenvalues and Eigenvectors of the Pendulum

For the overdamped case, let the constants in the system be $m = 0.1$, $l = 1$, $b = 1$, and $g = 9.8$. The resulting model is

$$\begin{bmatrix} \dot{\theta} \\ \ddot{\theta} \end{bmatrix} = \begin{bmatrix} 0 & 1 \\ -9.8 & -10 \end{bmatrix} \begin{bmatrix} \theta \\ \dot{\theta} \end{bmatrix} \tag{2.165}$$

The eigenvalues are found using

$$|\lambda I - A| = 0$$

$$\left| \begin{bmatrix} \lambda & 0 \\ 0 & \lambda \end{bmatrix} - \begin{bmatrix} 0 & 1 \\ -9.8 & -10 \end{bmatrix} \right| = 0$$

$$\begin{vmatrix} \lambda & -1 \\ 9.8 & \lambda + 10 \end{vmatrix} = 0 \tag{2.166}$$

$$\lambda(\lambda + 10) + 9.8 = 0$$

Solving (2.166) gives the two distinct real eigenvalues for the system.

$$\lambda_1 = -1.1013$$
$$\lambda_2 = -8.8987 \tag{2.167}$$

The eigenvectors of the system are found by solving

$$\left(\begin{bmatrix} 0 & 1 \\ -9.8 & -10 \end{bmatrix} - \lambda \begin{bmatrix} 1 & 0 \\ 0 & 1 \end{bmatrix} \right) \begin{bmatrix} \widehat{x}_1 \\ \widehat{x}_2 \end{bmatrix} = 0$$

$$\begin{bmatrix} -\lambda & 1 \\ -9.8 & -10 - \lambda \end{bmatrix} \begin{bmatrix} \widehat{x}_1 \\ \widehat{x}_2 \end{bmatrix} = 0 \tag{2.168}$$

for each eigenvalue. Solving these systems of equations for each λ gives

$$
\begin{aligned}
\widehat{x}_{\lambda_1} &= \begin{bmatrix} 1 \\ \lambda_1 \end{bmatrix} = \begin{bmatrix} 1 \\ -1.1013 \end{bmatrix} \\
\widehat{x}_{\lambda_2} &= \begin{bmatrix} 1 \\ \lambda_2 \end{bmatrix} = \begin{bmatrix} 1 \\ -8.8987 \end{bmatrix}
\end{aligned}
\tag{2.169}
$$

Often the eigenvectors are normalized so that they have unit length. Using this convention, the eigenvectors become

$$
\begin{aligned}
\widehat{x}_{\lambda_1} &= \begin{bmatrix} \dfrac{1}{\sqrt{1+\lambda_1^2}} \\ \dfrac{\lambda_1}{\sqrt{1+\lambda_1^2}} \end{bmatrix} = \begin{bmatrix} 0.6722 \\ -0.7403 \end{bmatrix} \\
\widehat{x}_{\lambda_2} &= \begin{bmatrix} \dfrac{1}{\sqrt{1+\lambda_2^2}} \\ \dfrac{\lambda_2}{\sqrt{1+\lambda_2^2}} \end{bmatrix} = \begin{bmatrix} 0.1117 \\ -0.9937 \end{bmatrix}
\end{aligned}
\tag{2.170}
$$

The code below shows the MATLAB commands to do the same eigenvalue calculation.

```
% pendulum_eig.m

close all
clear all

% Define the model parameters
m = 0.1;    % kg
l = 1;      % m
b = 1;      % kg*m^2/s
g = 9.8;    % m/s^2

A = [0, 1; -g/l, -b/(m*l^2)];
[x_hat, lambda] = eig(A)
```

The command for calculating eigenvalues and eigenvectors is `eig`. Using this command alone, `eig(A)` will return only the eigenvalues of square matrix `A`. To get the eigenvectors also, two output arguments must be specified.

```
[x_hat, lambda] = eig(A)
```

Both of the output variables `x_hat` and `lambda` are matrices that are the same size as `A`. The variable `lambda` is a diagonal matrix whose entries are the eigenvalues. Each column of `x_hat` is the eigenvector corresponding to the eigenvalue in the same column of `lambda`. The following values are output.

```
x_hat =
     0.6722   -0.1117
    -0.7403    0.9937
lambda =
    -1.1013         0
          0   -8.8987
```

Thus, the eigenvector associated with eigenvalue -1.1013 is $[0.6722, -0.7403]^T$ and with eigenvalue -8.8987 is $[-0.1117, 0.9937]^T$. These values match those given in (2.170) except for a sign change in the second eigenvector $[0.1117, -0.9937]^T$. This sign change does not change the eigenvector because it still defines the same line, just in the opposite direction.

What do these eigenvalues and eigenvectors mean physically? The phase plot provides insight into this question. In Figure 2.48, the phase plot shows the pendulum starting from an initial angle of 1 radian with zero initial angular velocity. At the moment the pendulum is released, the vector field is $[0, -9.8]$, which is obtained by plugging the initial conditions into (2.165). Recall that this vector field represents $[\dot{\theta}, \ddot{\theta}]^T$, so it gives the direction of change in the phase plot. Physically, it means the pendulum is experiencing no position change and a velocity change in the negative direction at that instant. As the phase plot progresses, we see that the pendulum stabilizes to the origin with no oscillations (as can been from the fact that the angle never becomes negative), as expected for an overdamped pendulum.

Now consider the phase plot of the pendulum with initial condition at $(1, -1.1013)$, a scalar multiple of one of the eigenvectors. The corresponding phase plot is shown in Figure 2.49. As shown in the plot, the vector field corresponding to the initial condition is along the same line as the initial condition vector but in the opposite direction. From this point, one can see that the negative eigenvalue means that an initial condition

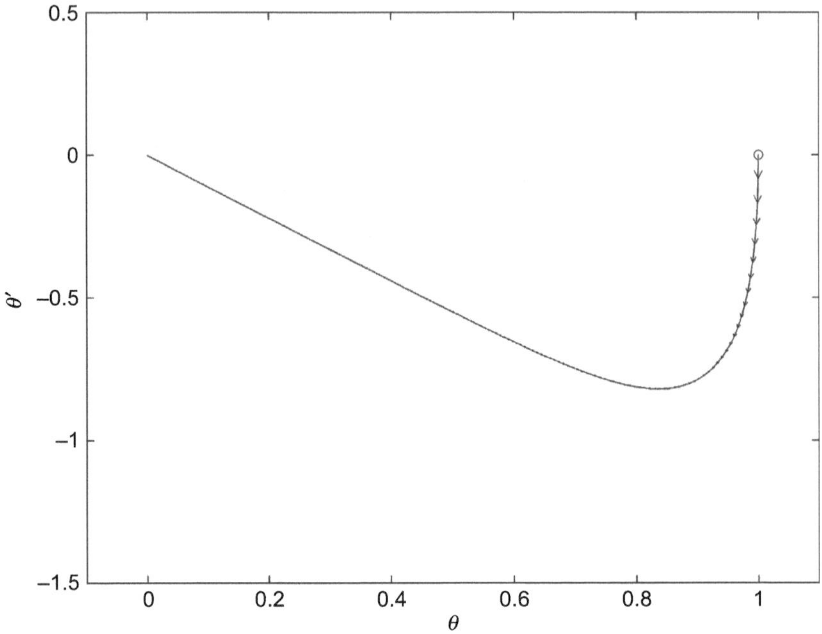

Figure 2.48 Phase plot of overdamped pendulum with initial condition (1, 0).

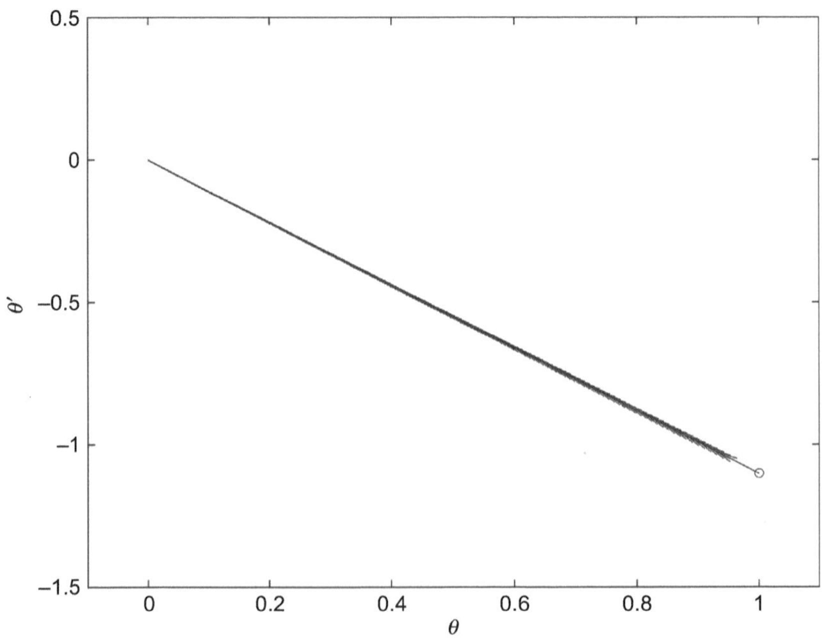

Figure 2.49 Phase plot of overdamped pendulum with initial condition (1, −1.1013).

that lies along the eigenvector will be "pulled" directly back to the origin. Alternatively, if the eigenvalue is positive, any initial condition that lies along the eigenvector will be "pushed" away from the origin. From this point of view, we can see why stable systems have negative eigenvalues associated with them, a characteristic that is discussed more in Chapter 3.

Let us now consider an underdamped pendulum. For this case, the constants in the system are $m = 0.1$, $l = 1$, $b = 0.1$, and $g = 9.8$. Using the same code as above with the new values, the eigenvalues are found to be

```
x_hat =
  -0.0486 - 0.3004i -0.0486 + 0.3004i
   0.9526             0.9526
lambda =
  -0.5000 + 3.0903i         0
        0            -0.5000 - 3.0903i
```

In this case, the eigenvalues are complex conjugates.[9]

$$\begin{aligned} \lambda_1 &= -0.5 + 3.0903j \\ \lambda_2 &= -0.5 - 3.0903j \end{aligned} \tag{2.171}$$

The corresponding eigenvectors are

$$\begin{aligned} \widehat{x}_{\lambda_1} &= \begin{bmatrix} -0.0486 - 0.3004j \\ 0.9526 \end{bmatrix} \\ \widehat{x}_{\lambda_2} &= \begin{bmatrix} -0.0486 + 0.3004j \\ 0.9526 \end{bmatrix} \end{aligned} \tag{2.172}$$

which also contain complex elements.

The same question may be asked as before: What do these eigenvalues and eigenvectors mean physically? Unfortunately, we cannot use the geometric interpretation as we did in the case of real eigenvalues and eigenvectors. As eloquently stated by Hirsch, Smale, and Devaney (2013), "Now in general it is not polite to hand someone a complex solution to a real system of differential equations..."

[9] Here MATLAB uses i to denote $\sqrt{-1}$, but it also recognizes j. The convention in this book is to use j because the author has an electrical engineering background.

As shown in Section 3.2.4, eigenvalues and eigenvectors play an important role in the analytical solution for a linear system. A solution can be expressed as

$$x(t) = e^{\lambda t}\,\widehat{x} \tag{2.173}$$

where λ is an eigenvalue, and $\widehat{x}$ is a eigenvector. With the help of Euler's formula,

$$e^{j\theta} = \cos\theta + j\sin\theta \tag{2.174}$$

the solution (2.175) becomes

$$\begin{aligned} x(t) &= e^{\lambda t}\,\widehat{x} \\ &= e^{\alpha t}(\cos(\beta t) + j\sin(\beta t))\widehat{x} \end{aligned}$$

if the complex eigenvalue is $\lambda = \alpha + j\beta$. Taking either the real or imaginary part of $x(t)$, is also a solution. Although it may seem complex (pun intended!), the imaginary part of the expression really is just an indication that the solution is oscillatory in nature. The real part tells us whether the solution grows or shrinks (as we saw previously with the initial condition being pulled toward or pushed away from the origin). With the real and imaginary parts taken together, the phase plot will form a spiral going toward the origin (for negative real parts of λ) or away from the origin (for positive real parts of λ) with the magnitude of the real part indicating how fast it moves. The phase plot of the underdamped pendulum is shown in Figure 2.50.

2.4.6 Singular Value Decomposition

The idea of singular value decomposition (SVD) in a sense can be thought of as an extension of eigenvalues. First, consider the idea of eigenvalue decomposition. Let Λ be the diagonal matrix of eigenvalues, that is,

$$\Lambda = \begin{bmatrix} \lambda_1 & 0 & 0 & 0 \\ 0 & \lambda_2 & 0 & 0 \\ 0 & 0 & \lambda_3 & 0 \\ 0 & 0 & 0 & \lambda_4 \end{bmatrix} \tag{2.175}$$

for a fourth-order system. We can then rewrite the eigenvector expression in (2.159) in matrix form as

$$A[\widehat{x}_{\lambda_1} \quad \widehat{x}_{\lambda_2} \quad \widehat{x}_{\lambda_3} \quad \widehat{x}_{\lambda_4}] = \Lambda[\widehat{x}_{\lambda_1} \quad \widehat{x}_{\lambda_2} \quad \widehat{x}_{\lambda_3} \quad \widehat{x}_{\lambda_4}] \tag{2.176}$$

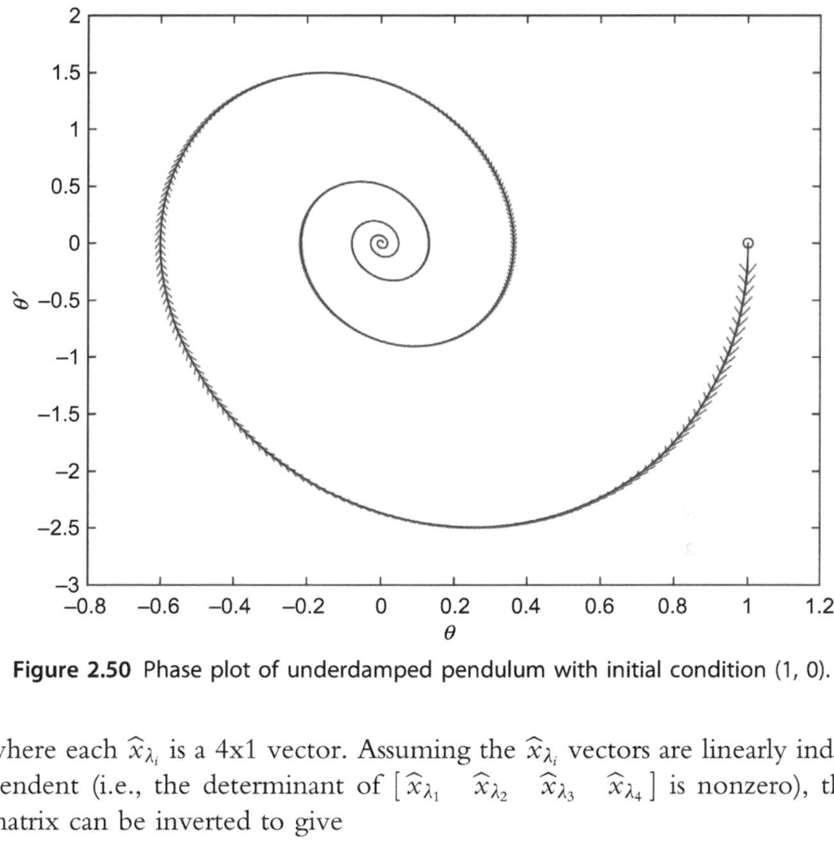

Figure 2.50 Phase plot of underdamped pendulum with initial condition (1, 0).

where each $\widehat{x}_{\lambda_i}$ is a 4x1 vector. Assuming the $\widehat{x}_{\lambda_i}$ vectors are linearly independent (i.e., the determinant of $[\,\widehat{x}_{\lambda_1} \quad \widehat{x}_{\lambda_2} \quad \widehat{x}_{\lambda_3} \quad \widehat{x}_{\lambda_4}\,]$ is nonzero), the matrix can be inverted to give

$$A = [\,\widehat{x}_{\lambda_1} \quad \widehat{x}_{\lambda_2} \quad \widehat{x}_{\lambda_3} \quad \widehat{x}_{\lambda_4}\,]^{-1}\Lambda[\,\widehat{x}_{\lambda_1} \quad \widehat{x}_{\lambda_2} \quad \widehat{x}_{\lambda_3} \quad \widehat{x}_{\lambda_4}\,] \tag{2.177}$$

which is known as the **eigenvalue decomposition** of A.

Now extending this idea to nonsquare matrices, suppose A is an mxn matrix. Then the **SVD** of A is

$$A = U\Sigma V^T \tag{2.178}$$

where U and V are orthonormal matrices, that is, $U^{-1} = U^T$ and $V^{-1} = V^T$, and Σ is of the form

$$\Sigma = \begin{bmatrix} \sigma_1 & \cdots & 0 & 0 & \cdots & 0 \\ \vdots & \ddots & \vdots & \vdots & \ddots & \vdots \\ 0 & \cdots & \sigma_r & 0 & \cdots & 0 \\ 0 & \cdots & 0 & 0 & \cdots & 0 \end{bmatrix} \tag{2.179}$$

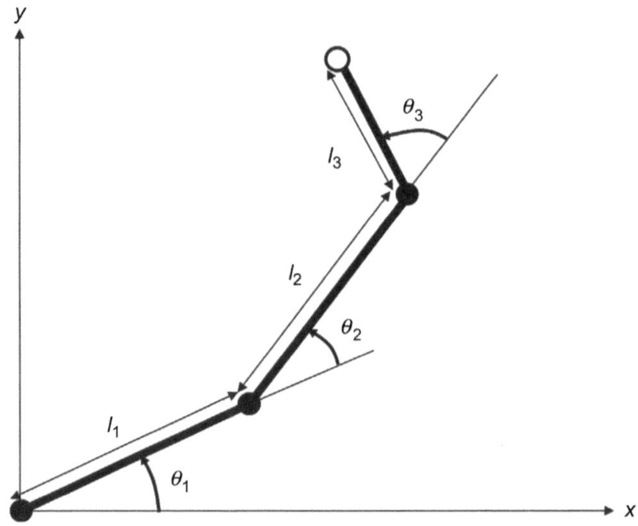

Figure 2.51 The three-link robotic arm restricted to movement in the *xy* plane.

and contains within it a diagonal matrix. The matrices are chosen so that the nonzero elements of Σ are positive and are arranged in decreasing value along the diagonal so that $\sigma_1 \geq \sigma_2 \geq \ldots \geq \sigma_r \geq 0$. These nonzero entries of Σ are the **singular values** of A.

There are many applications of SVD, including determining the rank of a matrix, solving linear equations, determining an orthonormal basis for A, computing the pseudoinverse of A, and computing projection operators, all of which have useful application in studying systems. Rather than discussing each use in isolation, the concepts associated with SVD will be demonstrated by working through an example of a robotic arm.[10] The SVD will have two main uses in this example.

1. To find a solution to the inverse kinematic problem using the pseudoinverse
2. To find the manipulability ellipses that show how easily the end effector can move in each direction

Figure 2.51 shows a three-link robotic arm confined to move in the xy plane. The three links have lengths l_1, l_2, and l_3, respectively. The angle of the first link is measured counterclockwise with respect to the x-axis. The

[10] This example is adopted from Bay (1999).

angles of subsequent links are measured counterclockwise with respect to the previous link.

The position of the end point of the robotic arm (commonly called the end effector) is given by

$$\begin{aligned} x &= l_1 \cos(\theta_1) + l_2 \cos(\theta_1 + \theta_2) + l_3 \cos(\theta_1 + \theta_2 + \theta_3) \\ y &= l_1 \sin(\theta_1) + l_2 \sin(\theta_1 + \theta_2) + l_3 \sin(\theta_1 + \theta_2 + \theta_3) \end{aligned} \tag{2.180}$$

The velocity of the end effector is then

$$\begin{aligned} \dot{x} &= -l_1 \sin(\theta_1)\dot{\theta}_1 - l_2 \sin(\theta_1 + \theta_2)(\dot{\theta}_1 + \dot{\theta}_2) - l_3 \sin(\theta_1 + \theta_2 + \theta_3)(\dot{\theta}_1 + \dot{\theta}_2 + \dot{\theta}_3) \\ \dot{y} &= l_1 \cos(\theta_1)\dot{\theta}_1 + l_2 \cos(\theta_1 + \theta_2)(\dot{\theta}_1 + \dot{\theta}_2) + l_3 \cos(\theta_1 + \theta_2 + \theta_3)(\dot{\theta}_1 + \dot{\theta}_2 + \dot{\theta}_3) \end{aligned} \tag{2.181}$$

Rearranging (2.181) in matrix form gives

$$\begin{bmatrix} \dot{x} \\ \dot{y} \end{bmatrix} = \begin{bmatrix} \begin{array}{c} -l_1 \sin(\theta_1) - l_2 \sin(\theta_1 + \theta_2) \\ -l_3 \sin(\theta_1 + \theta_2 + \theta_3) \end{array} & \begin{array}{c} -l_2 \sin(\theta_1 + \theta_2) \\ -l_3 \sin(\theta_1 + \theta_2 + \theta_3) \end{array} & -l_3 \sin(\theta_1 + \theta_2 + \theta_3) \\ \begin{array}{c} l_1 \cos(\theta_1) + l_2 \cos(\theta_1 + \theta_2) \\ +l_3 \cos(\theta_1 + \theta_2 + \theta_3) \end{array} & \begin{array}{c} l_2 \cos(\theta_1 + \theta_2) \\ +l_3 \cos(\theta_1 + \theta_2 + \theta_3) \end{array} & l_3 \cos(\theta_1 + \theta_2 + \theta_3) \end{bmatrix} \begin{bmatrix} \dot{\theta}_1 \\ \dot{\theta}_2 \\ \dot{\theta}_3 \end{bmatrix} \tag{2.182}$$

The 2 × 3 matrix in (2.182) is known as the **Jacobian**, denoted by $J(\Theta)$, because its elements are

$$J(\Theta) = \begin{bmatrix} \dfrac{\partial x}{\partial \theta_1} & \dfrac{\partial x}{\partial \theta_2} & \dfrac{\partial x}{\partial \theta_3} \\ \dfrac{\partial y}{\partial \theta_1} & \dfrac{\partial y}{\partial \theta_2} & \dfrac{\partial y}{\partial \theta_3} \end{bmatrix} \tag{2.183}$$

The expression in (2.182) describes the forward kinematics of the robotic arm. That is, given the joint angles and their angular velocities, we can determine how the position of the end effector moves. Designers of robots are often concerned with the more difficult problem of inverse kinematics in which the task is to determine how to move the joint angles to achieve some desired movement of the end effector. In this situation, the problem reduces to viewing (2.182) as a system of two equations with three unknowns. In general, there is more than one solution to this problem, and the SVD of the Jacobian will help find an answer.

Solving (2.182) for the angles gives

$$\begin{bmatrix} \dot{\theta}_1 \\ \dot{\theta}_2 \\ \dot{\theta}_3 \end{bmatrix} = J^{+}(\Theta) \begin{bmatrix} \dot{x} \\ \dot{y} \end{bmatrix} \tag{2.184}$$

where the $^+$ operator denotes the pseudoinverse. According to the theory of simultaneous equations, using the pseudoinverse gives the minimum norm solution. In this case, that means the solution minimizes $\frac{1}{2}(\dot{\theta}_1^2 + \dot{\theta}_2^2 + \dot{\theta}_3^2)$. Using the SVD of $J(\Theta)$, it can be expressed as

$$J(\Theta) = U \Sigma V^T \tag{2.185}$$

and then (2.184) becomes

$$\begin{bmatrix} \dot{\theta}_1 \\ \dot{\theta}_2 \\ \dot{\theta}_3 \end{bmatrix} = V \Sigma^{+} U^T \begin{bmatrix} \dot{x} \\ \dot{y} \end{bmatrix} \tag{2.186}$$

2.4.6.1 MATLAB Example: Inverse Kinematics of the Robotic Arm

This example shows how to solve the inverse kinematic problem using SVD in MATLAB and plots the results. In the problem, the desired movement of the end effector is given in terms of $\dot{x}$ and $\dot{y}$. The code then determines the necessary angular velocities to achieve that end-effector movement. The angles are updated according to these angular velocities to get the angles at the next sampling time. The new angles are then used in the next iteration of the solution. The code below is the basic program for simulating the robotic arm movement.

```
% inverse_kinematics.m

close all
clear all

% Set the lengths of each link
l1 = 1;
l2 = 1.5;
l3 = 0.5;

% Sampling time for updating
T = 0.01;
```

```
% Initial conditions
theta1 = 45*pi/180;
theta2 = 60*pi/180;
theta3 = -20*pi/180;
theta = [theta1;theta2;theta3];

% Desired trajectory
x_dot = 5;
y_dot = -5;

% Plot the initial position
figure
hold on
plot_arm(l1,l2,l3,theta,'bo')

% Update the plot for several iterations
for k = 1:10

    % Define the Jacobian matrix
    J(1,1) = -l1*sin(theta(1)) - l2*sin(theta(1)+theta(2)) -
l3*sin(theta(1)+theta(2)+theta(3));
    J(1,2) = J(1,1) + l1*sin(theta(1));
    J(1,3) = J(1,2) + l2*sin(theta(1)+theta(2));
    J(2,1) = l1*cos(theta(1)) + l2*cos(theta(1)+theta(2)) +
l3*cos(theta(1)+theta(2)+theta(3));
    J(2,2) = J(2,1) - l1*cos(theta(1));
    J(2,3) = J(2,2) - l2*cos(theta(1)+theta(2));

    % Perform singular value decomposition
     [U,S,V] = svd(J);

    % Determine angular velocities
    [theta_dot] = V*pinv(S)*U'*[x_dot;y_dot];

    % Update angles
    theta = theta+theta_dot*T;
    plot_arm(l1,l2,l3,theta,'b*')

end

% Format plot
xlabel('x')
ylabel('y')
axis equal
axis([-(l1+l2+l3)/2 (l1+l2+l3)/2 0 (l1+l2+l3)])
```

After the initial setting of the arm lengths, the sampling time is set to 0.01 s. The sampling time is needed because the angles have to be updated for the next iteration of the solution. This method allows us to see the progression of the arm over time.

Next, initial conditions of the arm position are set so the arm angles are 45, 60, and −20 degrees. The desired movement of the end effector is to move down and right at a 45-degree angle, so $\dot{x} = 5$ and $\dot{y} = -5$. Using SI units, these values are in meters per second. Although 5 m/s is a large velocity for a robotic arm to achieve, these values are used for illustrative purposes in the simulation because it allows the movements to be clear.

The next step in the code is to plot the initial position of the arm in a figure. This plot is accomplished by calling the `plot_arm` function, which is shown below. The arm is drawn as three vectors using the `quiver` command. The joints are drawn as individual points using the specified marker, which is passed to the function. For the initial condition, the joint positions are drawn as circles. Subsequent joint positions are drawn as asterisks.

```
function plot_arm(l1,l2,l3,theta,marker)

x1 = l1*cos(theta(1));
y1 = l1*sin(theta(1));
x2 = x1+l2*cos(theta(1)+theta(2));
y2 = y1+l2*sin(theta(1)+theta(2));
x3 = x2+l3*cos(theta(1)+theta(2)+theta(3));
y3 = y2+l3*sin(theta(1)+theta(2)+theta(3));

quiver(0,0,l1*cos(theta(1)),l1*sin(theta(1)),1,marker)
quiver(x1,y1,l2*cos(theta(1)+theta(2)),l2*sin(theta(1)+
theta(2)),1,marker)
quiver(x2,y2,l3*cos(theta(1)+theta(2)+theta(3)),l3*sin(theta(1)+
theta(2)+theta(3)),1,marker)
plot(x3,y3,marker)
```

After the initial plot, the mathematical solution of the inverse kinematic problem begins. In the code, 10 positions are determined at 0.01-s intervals.

The first step is to define the Jacobian matrix $J(\Theta)$. This step needs to be done each time through the loop because as the angles change, the values in $J(\Theta)$ need to be updated. Then the SVD is performed on $J(\Theta)$ using the `svd` command, returning the three matrices specified in (2.185). After these matrices are determined, the desired angular velocities are found by implementing (2.186) using matrix multiplication. The `'` operator in MATLAB takes the transpose of a matrix, and the `pinv` command takes the pseudoinverse of a matrix.

After the angular velocities are determined, the new angles are calculated using the first-order approximation

$$\theta(t+\Delta t) \approx \theta(t) + \frac{d\theta}{dt}\Delta t$$

or in MATLAB code,

```
theta = theta+theta_dot*T;
```

where Δt and `T` are the sampling time of 0.01. Then the plot is updated to illustrate the new arm position, and the process is repeated.

Notice that the actual mathematical solution occurs in two lines.

```
% Perform singular value decomposition
[U,S,V] = svd(J);

% Determine angular velocities
[theta_dot] = V*pinv(S)*U'*[x_dot;y_dot];
```

The majority of the code is for other peripheral, but important, tasks such as setting constants, updating variables, and plotting the results.

The plot resulting from the above code is shown in Figure 2.52. As was desired, the end effector moves down and right at a 45-degree angle from its initial position.

Figure 2.53 shows the result of the same code using the same initial condition but a different desired end effector movement. In this case, the values are $\dot{x} = 5$ and $\dot{y} = 5$.

At first, the arm is moving as desired, and the end effector is moving up and right at a 45-degree angle. But on the last iteration, something strange and interesting happens. It starts to pull back in. The reason for this behavior is that the robot is nearly completely stretched out ($\theta_2 = 0$ and $\theta_3 = 0$) and near its kinematic singularity. In this configuration, the arm cannot move outward as quickly as it can in other directions. We can see

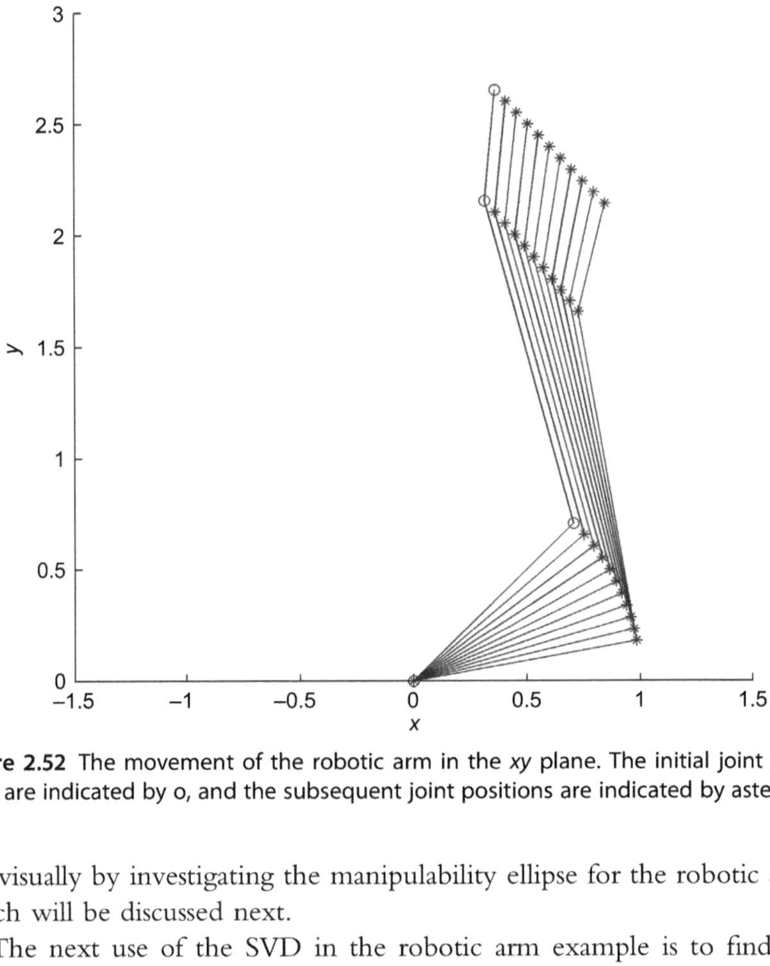

Figure 2.52 The movement of the robotic arm in the *xy* plane. The initial joint positions are indicated by o, and the subsequent joint positions are indicated by asterisks.

this visually by investigating the manipulability ellipse for the robotic arm, which will be discussed next.

The next use of the SVD in the robotic arm example is to find the manipulability ellipses. A **manipulability ellipse** is a means to visualize in what directions the end effector can move when the arm is in a particular configuration.

As a starting point in the discussion, consider the robotic arm stretched out along the x-axis so that $\theta_1 = \theta_2 = \theta_3 = 0$ as in Figure 2.54. Then according to (2.182), $\dot{x} = 0$ and therefore there can be no movement in the radial direction. This situation is known as a **kinematic singularity** in the robot's movement. Mathematically, singularities can be found by investigating the Jacobian J and determining where JJ^T is not full rank, that is, the determinant of JJ^T is zero.

This singularity is present not only in the configuration shown in Figure 2.54 but is in fact the case whenever $\theta_2 = \theta_3 = 0$. The determinant

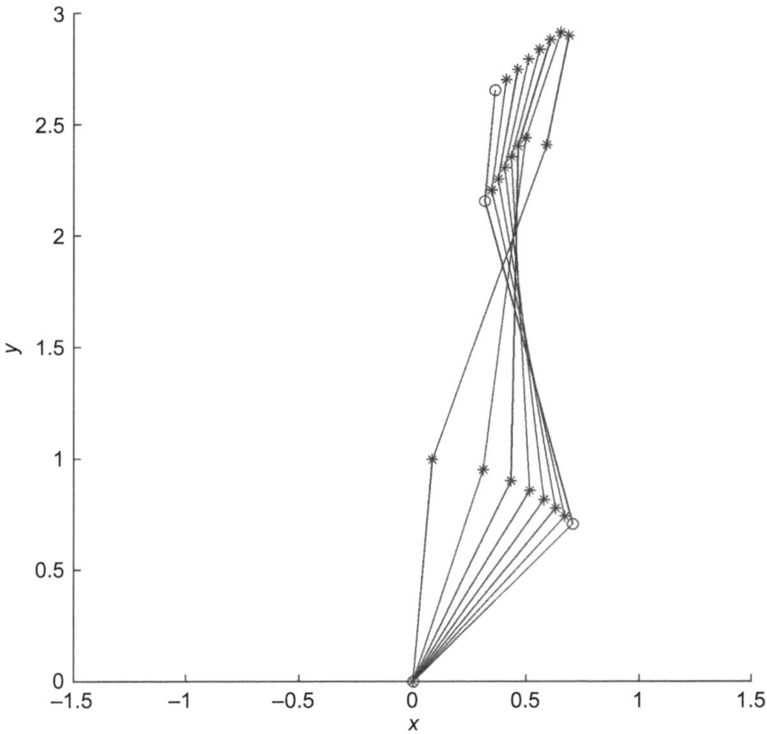

Figure 2.53 The robotic arm attempting to move toward its outstretched position.

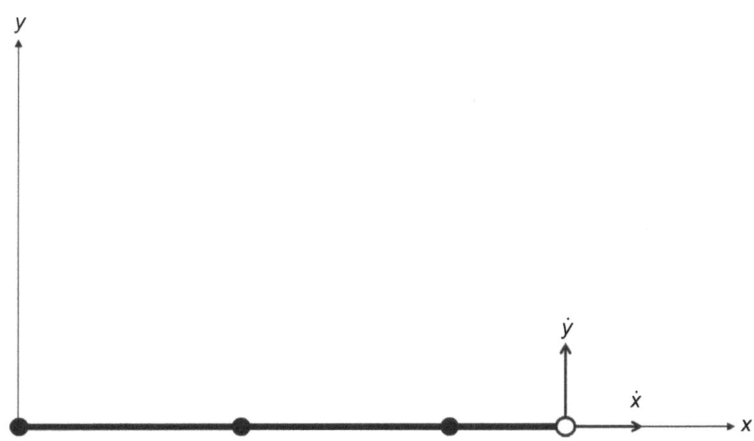

Figure 2.54 The robotic arm stretched out along the *x*-axis. From this configuration, the end effector cannot move along the *x*-axis.

of JJ^T is zero under these conditions. Thus, there can be no movement in the radial direction. This characteristic is intuitive in the outward direction (the arm cannot reach beyond its fully stretched position). However, in the inward direction, it is somewhat surprising, even counterintuitive to our everyday experience. If the reader has ever opened a bifold closet door, then the mathematics has seemingly been violated. To resolve this conflict, one must realize that the closet door and its track are never perfectly machined, and it allows some movement in the perpendicular direction. Any movement, even the smallest amount, brings the system away from the singularity and prevents it from getting "stuck" when trying to move in the radial direction.

We can now extend this idea to any end-effector position. Consider the two arm configurations shown in Figure 2.55. The arm in position B is "closer" to the singularity in the sense that it is more outstretched than the arm in position A. In position B, the end effector is relatively less capable of movement. We can characterize this idea in the manipulability ellipse.

The size and shape of the manipulability ellipse can be found using the SVD of J. The axis lengths of the ellipse are the singular values and the axis directions are the columns of U. At the singularities, the ellipse collapses to a line perpendicular to the arm.

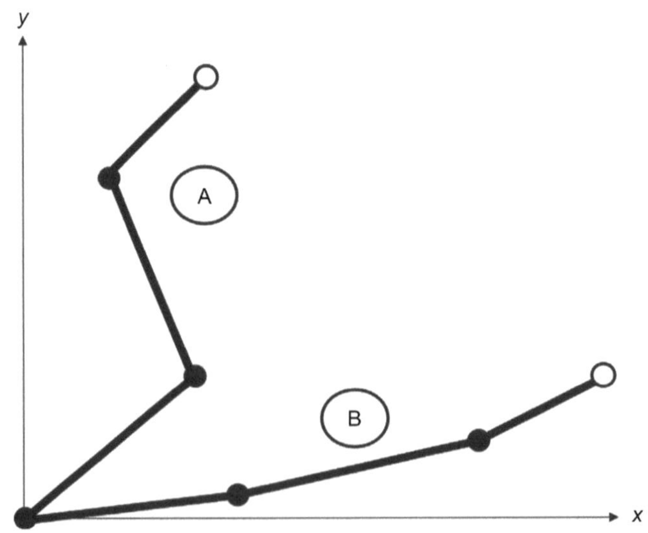

Figure 2.55 Two configurations of the robotic arm. The arm in position B is closer to kinematic singularity than the one in position A.

2.4.6.2 MATLAB Example: Manipulability Ellipse of the Robotic Arm

In this example, the manipulability ellipse is plotted three different configurations of the robot arm. The code below performs these operations.

```
% manipulability_ellipse.m

close all
clear all

% Set the lengths of each link
l1 = 1;
l2 = 1;
l3 = 1;

% Set arm position
theta1 = 15*pi/180;
theta2 = 100*pi/180;
theta3 = -150*pi/180;
theta = [theta1;theta2;theta3];

% Define the Jacobian matrix
J(1,1)    =    -l1*sin(theta(1))    -    l2*sin(theta(1)+theta(2))    -
l3*sin(theta(1)+theta(2)+theta(3));
J(1,2) = J(1,1) + l1*sin(theta(1));
J(1,3) = J(1,2) + l2*sin(theta(1)+theta(2));
J(2,1)    =    l1*cos(theta(1))    +    l2*cos(theta(1)+theta(2))    +
l3*cos(theta(1)+theta(2)+theta(3));
J(2,2) = J(2,1) - l1*cos(theta(1));
J(2,3) = J(2,2) - l2*cos(theta(1)+theta(2));

% Perform singular value decomposition
[U,S,V] = svd(J);

% Create the ellipse
sc = 0.3;
a = sc*(S(1,1));
b = sc*(S(2,2));
t = 0:0.01:2*pi;
x = a*cos(t);
y = b*sin(t);

% Apply rotation operator
A = U*[x;y];
x = A(1,:);
y = A(2,:);
```

```
% Apply translation operator
x = x + l1*cos(theta(1)) + l2*cos(theta(1)+theta(2)) +
l3*cos(theta(1)+theta(2)+theta(3));
y = y + l1*sin(theta(1)) + l2*sin(theta(1)+theta(2)) +
l3*sin(theta(1)+theta(2)+theta(3));

figure
hold on
plot_arm(l1,l2,l3,theta,'bo')
plot(x,y,':')

% Format plot
xlabel('x')
ylabel('y')
axis equal
axis([0 (l1+l2+l3) 0 (l1+l2+l3)])
```

The first section of the code is identical to the previous example and is where the link lengths, link angles, and Jacobian matrix are defined, and then the SVD of the Jacobian is found. After this, the lengths of the principle axes, typically denoted by a and b, are extracted from the SVD matrix `S`. A scale factor, `sc`, with a value of 0.3 is applied to the lengths so that the ellipse is small enough to fit in the figure window. Then the (x,y) points on the ellipse are generated using the formula

$$x = a\cos(t)$$
$$y = b\sin(t)$$

where $0 \leq t \leq 2\pi$.

Next, the rotation operator is applied to the (x, y) points of the ellipse. Because the columns of `U` are the directions of the ellipse's axes, simply multiplying `U` with the (x, y) points provides the appropriate rotation. Finally, the ellipse is translated so its center is the end-effector position. This translation is performed by adding the end-effector coordinates to each point on the ellipse.

After the ellipse is generated, the robotic arm is plotted as before and then the ellipse is added to the plot.

Figure 2.56 shows the manipulability ellipse for link angles $\theta_1 = 15$ degrees, $\theta_2 = 100$ degrees, and $\theta_3 = -150$ degrees. From the figure, it can be seen that the arm can move in all directions from this location, but it has more mobility along the y-axis compared with the x-axis. In contrast,

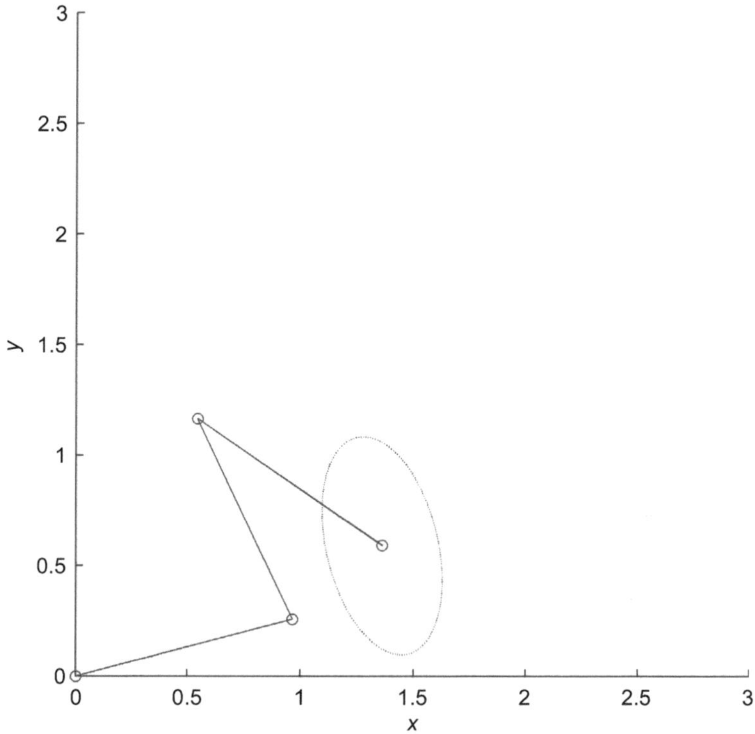

Figure 2.56 The manipulability ellipse for the arm when $\theta_1 = 15$ degrees, $\theta_2 = 100$ degrees, and $\theta_3 = -150$ degrees.

Figure 2.57 shows the arm with link angles set to $\theta_1 = 35$ degrees, $\theta_2 = 30$ degrees, and $\theta_3 = -15$ degrees. In this configuration, the arm is much more stretched out, and its manipulability ellipse is more oblong, reflecting the fact that the end effector cannot move radially outward as much as it can perpendicularly.

Finally, Figure 2.58 shows the case when the arm is fully stretched out with $\theta_1 = 50$ degrees, $\theta_2 = 0$ degrees, and $\theta_3 = 0$ degrees. In this configuration, the ellipse collapses to a line because the singular values of the Jacobian matrix are 3.7147 and 0. The corresponding directions of the ellipse axis are (from the columns of U) $[-0.7660\ 0.6428]^T$ and $[0.6428\ 0.7660]^T$. Thus, the major axis is 140 degrees $\left(= \tan^{-1}\left(\frac{0.6428}{-0.7660}\right)\right)$ with respect to the x-axis, and the minor axis is at 50 degrees. In this latter direction, the arm cannot move.

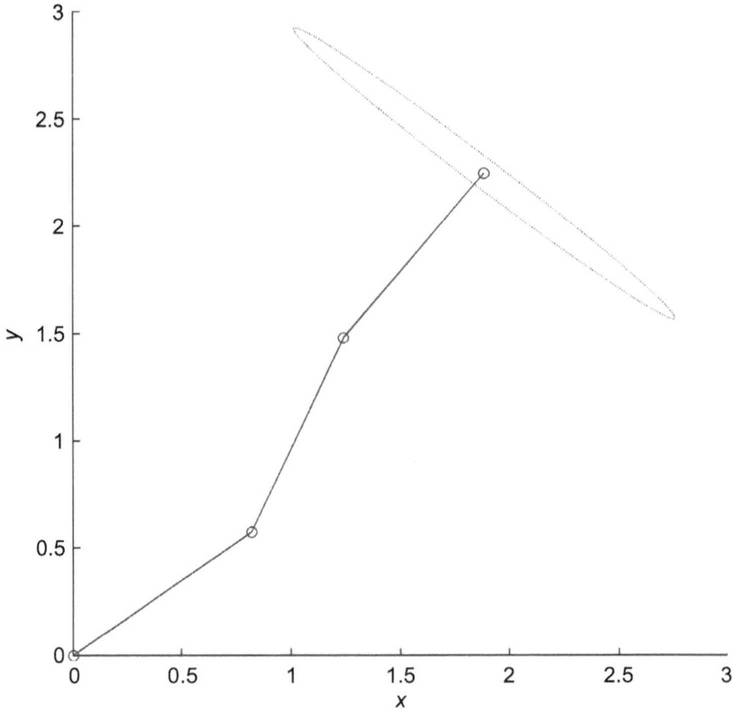

Figure 2.57 The manipulability ellipse for the arm when $\theta_1 = 35$ degrees, $\theta_2 = 30$ degrees, and $\theta_3 = -15$ degrees.

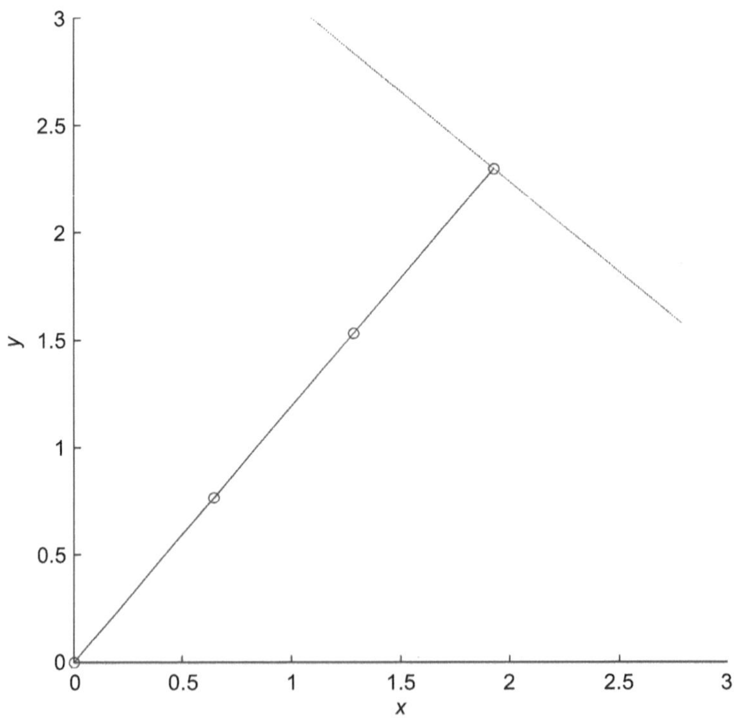

Figure 2.58 The manipulability ellipse for the arm when $\theta_1 = 50$ degrees, $\theta_2 = 0$ degrees, and $\theta_3 = 0$ degrees.

2.5 SYSTEM IDENTIFICATION

2.5.1 Overview

So far in this chapter, many models have been presented, and all of them have parameters. Parameters, in contrast to variables, are assumed to be constants (or very slowly varying) in the system. Bodies have mass, springs have stiffness, linkages have length, materials have thermal capacity, resistors have resistance, and the list continues. For any physical realization of a system, these parameters will have values, and these values must be accurately known to understand and be able to predict how the system will behave. The process of determining the parameter values is called **system identification**.

System identification is a field of study unto itself. Entire books have been written on the topic. One book that is considered required reading is by Ljung (1999). The results of the Ljung's work are the basis for the MATLAB System Identification Toolbox. The purpose of this section is to provide a brief overview of one method of system identification in the context of human balance as described in Goodworth et al. (2014).

2.5.2 Case Study: Human Balance Model

In this case study, we investigate a model for human balance and show how to estimate parameters in the model. Figure 2.59 shows the physical description of the model we are studying. Viewed from the front, the human body may be approximated as a two-link inverted pendulum. As the support surface rotates, a person standing on that surface generates two torques, one for the upper body T_U and one lower body T_L, so that balance is maintained. These torques depend on the angles of each body segment θ_U and θ_L, the pelvic angle θ_P, and the support surface angle θ_S. The goal is to determine *how* the torques depend on these angles to better understand how humans balance.

Figure 2.60 shows a model of how the simplified human balance system works. The upper part of the figure shows the mechanical modeling for the two links of the pendulum. The parameters in this part of the system (A's and J's) can be determined for each individual based on measurements such as weight, height, leg length, and so on. However, the parameters in the lower part of the system (K's, B's, G's, and τ's) cannot be measured directly and therefore must be calculated from the data.

For the study, frequency response data was collected from test subjects. To obtain the data, each test subject stood on a support surface that was

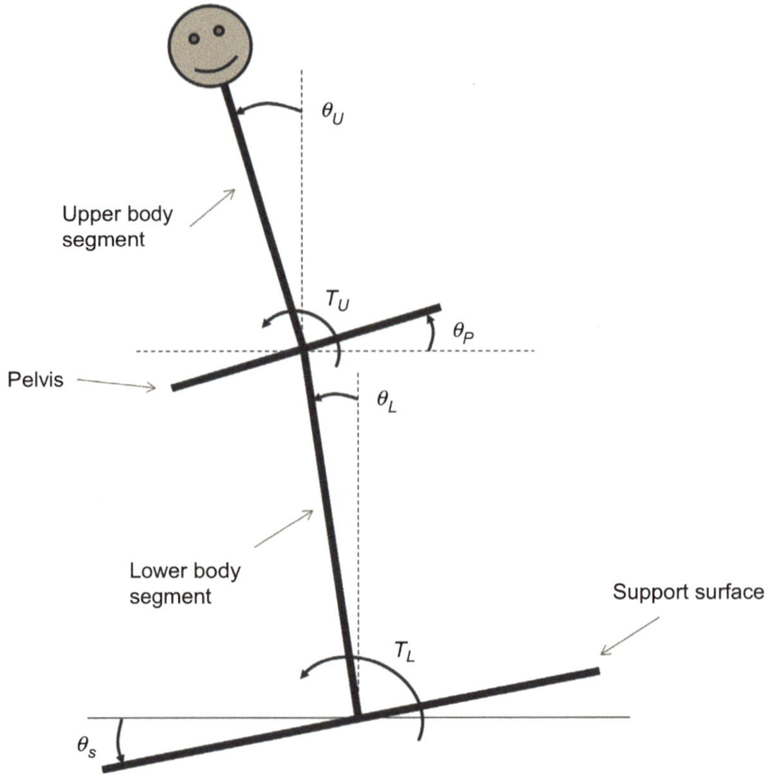

Figure 2.59 The model of frontal plane mechanics.

rotated. Upper and lower body sway was measured with angle sensors. To obtain useful data for system identification, it is important that the input signal (angular velocity of the support surface) be persistently exciting, meaning that it should contain enough frequencies to characterize the system behavior. There are many ways to meet this requirement, the simplest being to input sinusoids of varying frequency and measuring the output waveform. However, in this case, the human subjects are not given a deterministic signal such as a sinusoid because they may be able to predict the input, and the resulting data would not be representative of their balance reactions.

The input signal often applied for these types of studies is called a **pseudorandom ternary sequence** (PRTS). This signal is a randomly generated waveform that has three values ($0, +n, -n$). The pseudorandom term refers to the fact that the random sequence is generated by a computer algorithm, and so it is not truly random. An example of a PRTS waveform is shown in Figure 2.61, *A*. If this signal is input to the system to control the

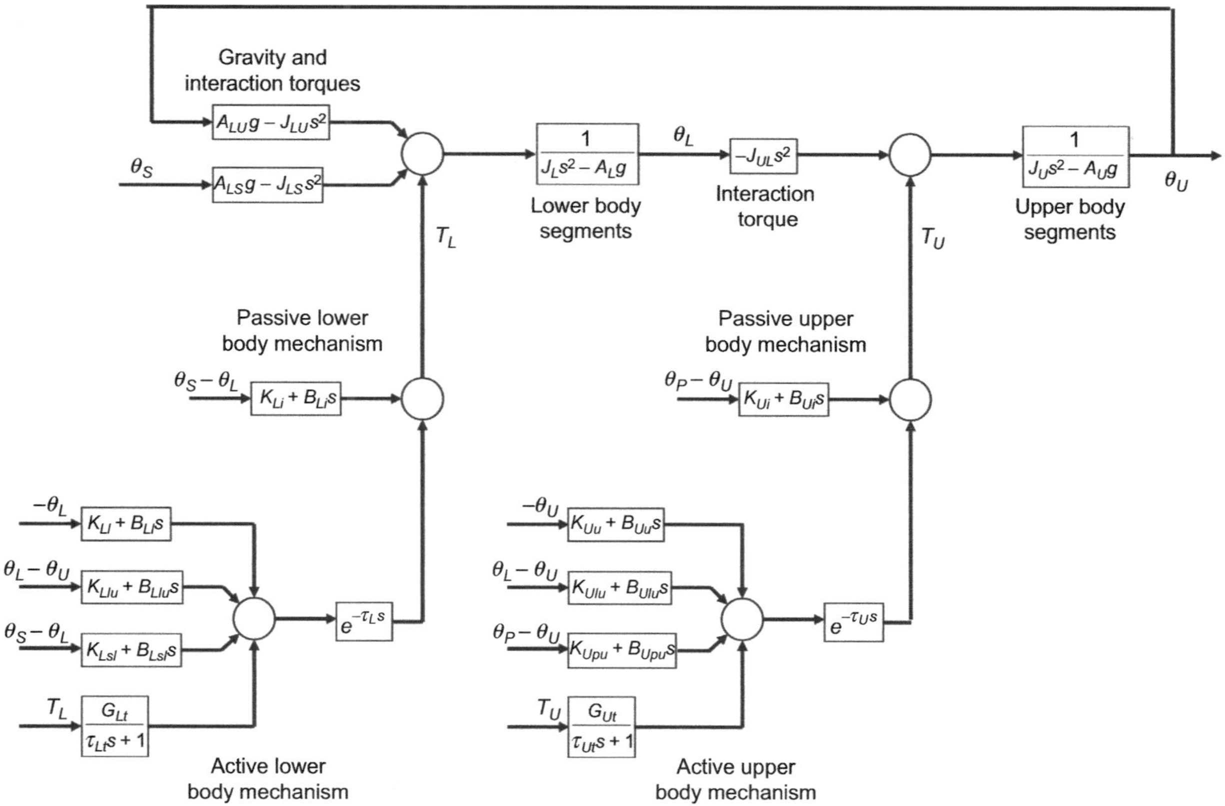

Figure 2.60 The block diagram of the balance system showing different contributing mechanisms.

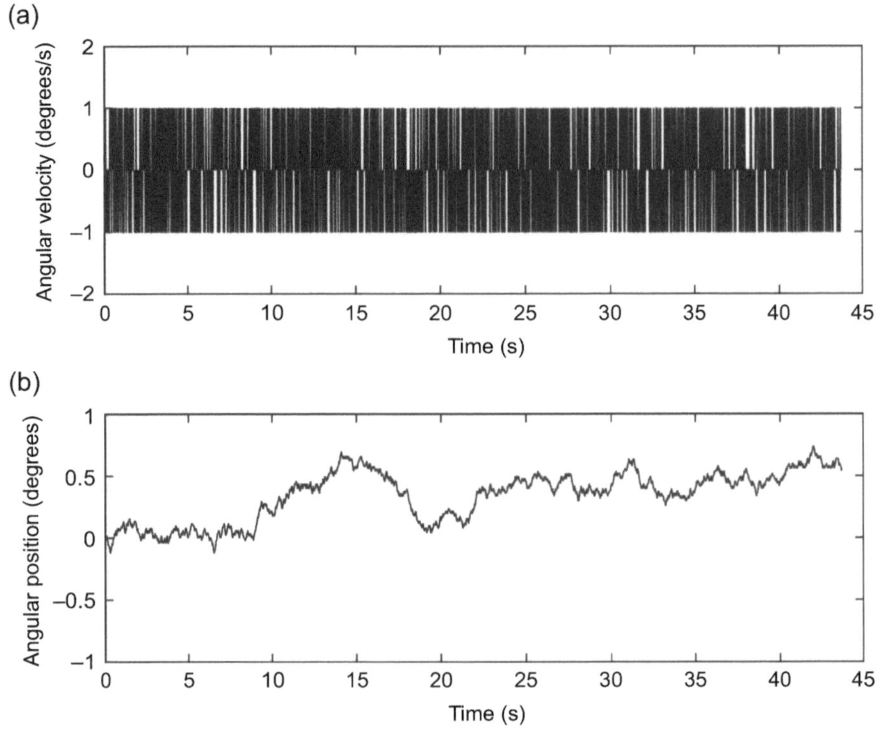

Figure 2.61 (a) A pseudorandom ternary sequence (PRTS) used to generate angular velocities for the platform. (b) The PRTS signal is integrated to obtain the angular position of the platform.

support surface velocity, the resulting angular position of that surface is shown in Figure 2.61, *B*. Advantages of using this waveform are that it (1) is random in nature and thus unpredictable for the test subjects and (2) has a flat power spectrum, meaning the frequency components have approximately equal amplitude across the bandwidth of interest. The power spectrum of the input signal is shown in Figure 2.62.

For each subject in the study, the upper body and lower body sway was measured while standing on the moving platform. The measured data were then used to formulate a frequency response function (FRF) for each subject, which was obtained from the ratio of discrete Fourier transforms of response signal to the platform stimulus signal. The FRFs are expressed as gain and phase values at various frequencies that represent the magnitude and timing of the subject's response to the stimulus. Figure 2.63 shows an example of FRF data. More extensive data can be found in Goodworth et al. (2014).

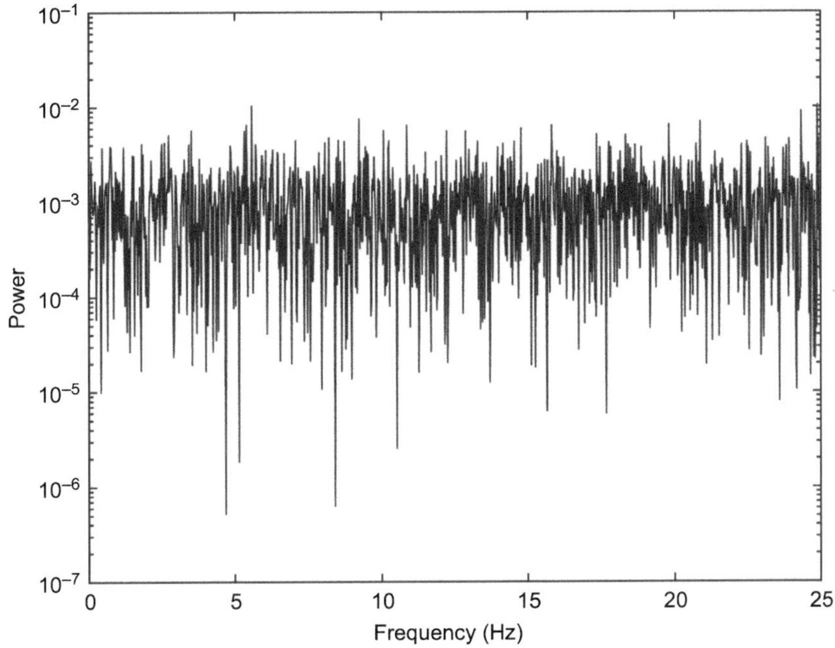

Figure 2.62 The power spectrum of the velocity signal.

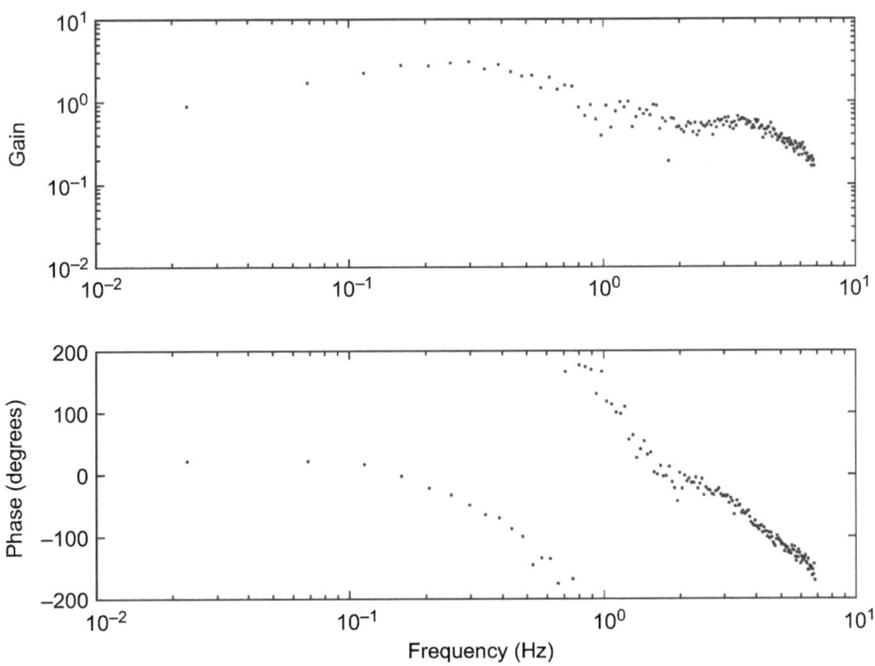

Figure 2.63 Example frequency response data.

With the FRFs for each subject and an average FRF based on all subjects, the parameters in the model can be estimated to obtain a model-predicted FRF. First the transfer functions for upper body and lower body sway with respect to the platform angle were calculated, $\frac{\theta_U(s)}{\theta_S(s)}$ and $\frac{\theta_L(s)}{\theta_S(s)}$ from the equations of motion of the body shown in Figure 2.59,

$$\begin{gathered}(J_L s^2 - A_L g)\theta_L(s) + (J_{LU}s^2 - A_{LU}g)\theta_U(s) + (J_{LS}s^2 - A_{LS}g)\theta_S(s) = RT_L(s) \\ J_{UL}s^2\theta_L(s) + (J_U s^2 - A_U g)\theta_U(s) = T_U(s)\end{gathered} \tag{2.187}$$

Then the model-predicted FRFs for upper body and lower body were derived from the transfer function with $s = j2\pi f$. The parameters in the analytical expression for the FRF were then chosen to minimize the error between $\frac{\theta_U(f)}{\theta_S(f)}$ and $\frac{\theta_L(f)}{\theta_S(f)}$ and the experimentally collected data points for upper and lower body sway.

This concludes our discussion on system modeling. We examined different types of modeling paradigms in depth, including equations of motion, transfer functions, and state-space models. While examining these models, we saw how they were used in applications such as car suspension and kinematic movement, bank accounts, human balance, heating systems, circuits, and robotic arms. Next we turn our attention from how systems are represented to their behavior. In the next chapter, we focus on some of the concepts we already encountered such as system trajectories, equilibrium points, and stability. We will formalize what they are and what they mean in terms of system behavior.

REFERENCES

Bay, J. (1999). *Fundamentals of Linear State Space Systems*. Boston: WCB/McGraw-Hill.

Cannon, R. (2003). *Dynamics of Physical Systems*. Mineola, NY: Dover Publications.

Dixon, J. (2007). *The Shock Absorber Handbook*. West Sussex, England: John Wiley & Sons.

Farlow, S. (1993). *Partial Differential Equations for Scientists and Engineers*. Mineola, NY: Dover Publications.

Franklin, G., Powell, J. D., & Emami-Naeini, A. (2010). *Feedback Control of Dynamic Systems* (6th ed.). Upper Saddle River, NJ: Pearson Higher Education.

Goebel, R., Sanfelice, R. G., & Teel, A. R. (2012). *Hybrid Dynamical Systems: Modeling, Stability, and Robustness*. Princeton, NJ: Princeton University Press.

Goodworth, A., Mellodge, P., & Peterka, R. J. (2014, August 1). Stance width changes how sensory feedback is used for multisegmental balance control. *Journal of Neurophysiology, 112*(3), 525–542.

Hirsch, M., Smale, S., & Devaney, R. L. (2013). *Differential Equations, Dynamical Systems, and an Introduction to Chaos* (3rd ed.). Waltham, MA: Academic Press.

Ljung, L. (1999). *System Identification: Theory for the User* (2nd ed.). Upper Saddle River, NJ: Prentice Hall PTR.
Moler, C. (2014, May 12). *Ordinary Differential Equation Suite*. Retrieved from Cleve's Corner: Cleve Moler on Mathematics and Computing: http://www.blogs.mathworks.com/cleve/2014/05/12/ordinary-differential-equation-suite/.
Peterka, R. (2003, March/April). Simplifying the Complexities of Maintaining Balance. *IEEE Engineering in Medicine and Biology, 22*(2), 63–68.
Yong, J. (2008). *Theory of Ground Vehicles* (4th ed.). Hoboken, NJ: John Wiley & Sons.

FURTHER READING

Fadali, M., & Visioli, A. (2009). *Digital Control Engineering: Analysis and Design*. Burlington, MA: Academic Press.
Jackson, L. (1990). *Signals, Systems, and Transforms*. Reading, MA: Addison-Wesley.
Kreyszig, E. (1993). *Advanced Engineering Mathematics* (7th ed.). New York: John Wiley & Sons.
Moler, C. (2008). *Numerical Computing with MATLAB* (2nd ed.). Philadelphia: Society for Industrial and Applied Mathematics.
Vaccaro, R. (1995). *Digital Control: A State-Space Approach*. New York: McGraw-Hill.
van der Schaft, A. J., & Schumacher, J. M. (2000). *An Introduction to Hybrid Dynamical Systems*. London: Springer-Verlag.

CHAPTER 3

Characteristics of Dynamical Systems

3.1 OVERVIEW

This chapter provides more detail about some general characteristics of dynamical systems, in particular their solutions, implications of equilibrium and stability, and the role of Lyapunov functions. These concepts apply to all types of dynamical systems, and we look at several different applications. In the next chapter, we focus on nonlinear systems and some characteristics peculiar to them.

In Section 3.2, we investigate the existence and uniqueness of solutions and why engineers should worry about them. We take an in-depth view of conditions for solutions to exist, apply the existence theorems to real-world systems, and look at what the results mean. Then the next step is taken, and we apply solution methods to linear systems. Section 3.3 examines certain solutions that are important in the study of dynamical systems: equilibria (or points where the dynamical system is no longer dynamic). We also study nullclines as a way of finding equilibrium points and dividing up a state-space according to system behavior. The important topic of stability is discussed in Section 3.4, and various definitions are given and discussed in the context of real-world systems. Finally, in Section 3.5, we investigate Lyapunov functions as a way to determine stability.

3.2 EXISTENCE AND UNIQUENESS OF SOLUTIONS: WHY IT MATTERS

The existence of solutions is very important when studying dynamical systems, particularly control systems, because there is no point in trying to solve a problem if there is no solution. For engineers, this first step may be easy to overlook because the main emphasis is on finding the solution and not stopping to consider if one actually exists. In fact, engineers, being the problem solvers that they are, may actually change a system to force a solution. Although this can be an appropriate path, it is important to realize what is happening and what the implications of such maneuvering may be.

A Practical Approach to Dynamical Systems for Engineers
ISBN 978-0-08-100202-5
http://dx.doi.org/10.1016/B978-0-08-100202-5.00003-6

In general, engineers focus on finding the solution to problems that mathematicians have previously proven are solvable. The differing points of view have been the topic of many jokes about engineers, mathematicians, and physicists. Here is one example from Mike Schuh's (n.d.) collection of engineer-mathematician-physicist jokes.[1]

An engineer, a mathematician, and a physicist are staying for the night in a hotel. Fortunately for this joke, a small fire breaks out in each room.

The physicist awakes, sees the fire, makes some careful observations, and on the back of the hotel's wine list does some quick calculations. Grabbing the fire extinguisher, he puts out the fire with one, short, well placed burst, and then crawls back into bed and goes back to sleep.

The engineer awakes, sees the fire, makes some careful observations, and on the back of the hotel's room service list (pizza menu) does some quick calculations. Grabbing the fire extinguisher (and adding a factor of safety of 5), he puts out the fire by hosing down the entire room several times over, and then crawls into his soggy bed and goes back to sleep.

The mathematician awakes, sees the fire, makes some careful observations, and on a blackboard installed in the room, does some quick calculations. Jubilant, he exclaims "A solution exists!", and crawls into his dry bed and goes back to sleep.

One of the strengths of mathematics is in its emphasis on explicit and precise definitions. Inspired by this practice, we will start with a definition.

3.2.1 What Is a Solution?

Before discussing the existence of solutions, we should first define what exactly is meant by a "solution." Let us address continuous-time and discrete-time systems separately.

For a discrete-time system, assume it is modeled by

$$x[n+k] = f(x[n+k-1], x[n+k-2], \ldots, x[n], n) \tag{3.1}$$

where x is a real-valued N-dimensional vector and the system has initial conditions

$$\begin{aligned} x[n_0] &= x_0 \\ x[n_0+1] &= x_1 \\ &\vdots \\ x[n_0+k-1] &= x_{k-1} \end{aligned} \tag{3.2}$$

[1] From http://www.farmdale.com/emp-jokes.shtml.

Note that this is a more general nonlinear and time-varying form than was given in Chapter 2, which focuses on linear systems. A **solution** to (3.1) is a sequence of numbers, starting with the given initial conditions $(x_0, x_1, \ldots, x_{k-1})$, and continuing on with $x_k, x_{k+1}, x_{k+2}, \ldots$ such that when these values are plugged into (3.1) the equality is satisfied.

For a continuous-time system, assume it is modeled by

$$\dot{x}(t) = f(t, x(t)) \tag{3.3}$$

where $x(t)$ is a real-valued N-dimensional vector and the system has initial conditions

$$x(t_0) = x_0 \tag{3.4}$$

As with the discrete-time case, here we are assuming a more general nonlinear and time-varying system than was given in Chapter 2. A **solution** to (3.3) is a function $\widehat{x}(t)$ such that $\dot{\widehat{x}}(t)$ is defined and when plugged into (3.3) satisfies the equality. Often the solution is defined over an interval of time, such as $[t_0, t_1]$, and that $\dot{\widehat{x}}(t) = f(t, \widehat{x}(t))$ for all $t \in [t_0, t_1]$. These solutions are also called **trajectories** or **orbits**. The phase plots shown in the previous chapter (such as in Figure 2.48) are examples of system trajectories.

The **uniqueness** of the solution means that the given solution is the only one that will work. There is *no other solution* that also satisfies the system equations. This condition guarantees that the system will move forward in time along the same path every time given the same starting point (initial condition).

3.2.2 Existence and Uniqueness Theorem

In many books on dynamical systems, a great deal of space is spent on theorems proving the existence and uniqueness of solutions to the equations that describe the systems.[2] The main result that is given is a theorem that provides sufficient conditions for the existence and uniqueness of the solution. Below is an example of such a theorem for continuous-time systems.[3]

Global Existence and Uniqueness: *Suppose $f(t, x(t))$ is piecewise continuous in t and satisfies the Lipschitz condition*

$$\|f(t,x) - f(t,y)\| \leq L\|x - y\| \tag{3.5}$$

[2] One book (Meiss, 2007) even provides three separate proofs of one version of the theorem.

[3] As stated in Khalil (1996).

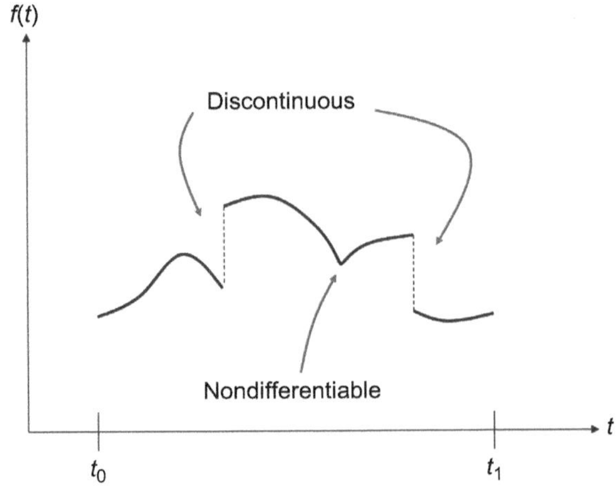

Figure 3.1 An example of a piecewise continuous function.

for all $x, y \in \Re^N$ *and for all* $t \in [t_0, t_1]$ *and also satisfies*

$$\|f(t, x_0)\| \leq h \tag{3.6}$$

for all $t \in [t_0, t_1]$. *Then the state equation* $\dot{x}(t) = f(t, x(t))$ *with* $x(t_0) = x_0$ *has a unique solution over* $[t_0, t_1]$.

Let us look at this theorem in detail and understand what it is saying. The first assumption is that f is piecewise continuous in t. Formally, from the definition of continuity, this means that for every $\varepsilon > 0$, there exists a $\delta > 0$ such that if $\|t - \hat{t}\| < \delta$ then $\|f(t, x(t)) - f(\hat{t}, x(\hat{t}))\| < \varepsilon$. **Continuity** means that the function f does not have jumps as it evolves in time. However, the piecewise modifier means that it can have jumps but only a finite number of them in any time interval. Figure 3.1 shows an example of a piecewise continuous function in time. The function $f(t)$ is **piecewise continuous** in the interval $[t_0, t_1]$ with two discontinuities. To illustrate the difference between continuous and differentiable, there is a point between the two discontinuities where f comes to a sharp point and thus is not differentiable, but it is continuous.

Notice that the continuous condition for f is with respect to t, not x. Instead the requirement on the x argument is the **Lipschitz condition**, which is stronger than continuity. In fact, continuity in x ensures that a solution exists,[4] but uniqueness requires the Lipschitz condition. This condition is stated in (3.5).

[4] A proof is given by Miller and Michel (1982).

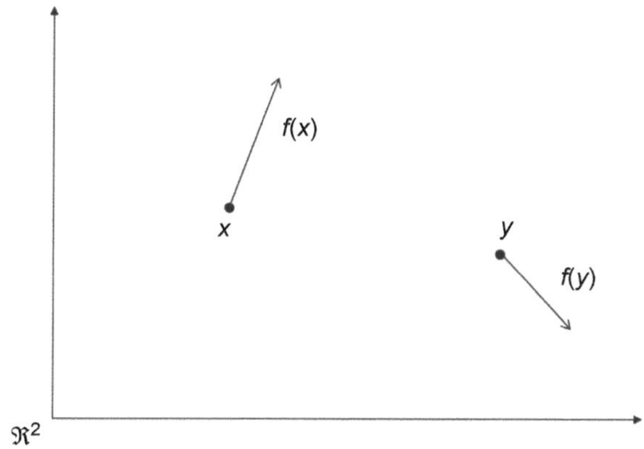

Figure 3.2 Two points, x and y, and their respective vector fields to illustrate the Lipschitz condition.

First, a note about the notation $||\bullet||$ (which also appeared in the definition of continuity). This is called the norm of a vector. It could be the Euclidean distance (as is often used in real-world systems), but it could also be a more general p-norm or an even more general form. A norm simply needs to satisfy three conditions:

1. $||x|| \geq 0$ for all $x \in \Re^N$ and $||x|| = 0$ if and only if $x = 0$.
2. $||x + y|| \leq ||x|| + ||y||$ for all $x, y \in \Re^N$.
3. $||\alpha x|| = |\alpha|\ ||x||$ for all $x \in \Re^N$ and $\alpha \in \Re$.

A p-norm is defined by

$$||x||_p = \left(x_1^p + x_2^p + \cdots + x_N^p\right)^{1/p} \tag{3.7}$$

and the Euclidean distance uses $p = 2$.

The Lipschitz condition in (3.5) lies somewhere between continuity and differentiability. That is, every function that is differentiable with continuous first derivative is a Lipschitz function, and every Lipschitz function is continuous.[5] This condition is illustrated for $N = 2$ in Figures 3.2 and 3.3. Two points x and y are chosen. From each of those points, the function f defines a vector field that indicates magnitude and direction of the instantaneous velocity shown by the arrows in Figure 3.2. In Figure 3.3, the distance between the points is shown as $||x - y||$, and the distance between the vector fields is shown as $||f(x) - f(y)||$. In this case, the norm is taken to be the Euclidean distance and is determined using vector subtraction. It should be noted that in this example, the system is time-invariant because

[5] Proof of these statements can be found in Meiss (2007).

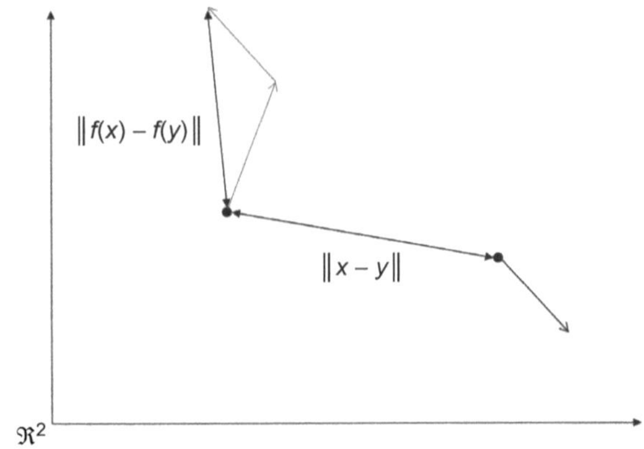

Figure 3.3 The distances between points x and y and the difference between their vector fields.

f does not depend explicitly on t. For time-varying systems (as described in the theorem earlier), the relationship needs to be true for every instant of time in the interval $[t_0, t_1]$.

The Lipschitz condition states that the distance between the velocities at two points is not too much bigger than the distance between the points themselves. The "not too much bigger" phrase is characterized by L, which is a constant real number.

The last condition of the theorem (3.6) states that the norm of the vector field emanating from the initial condition isn't too big. Here the "isn't too big" phrase is characterized by h. In other words, the system isn't moving too quickly from its initial state.

3.2.3 Application Examples

Let us explore two examples to see how this theorem works and why it's important.

Example 1: First consider the damped pendulum as discussed in Chapter 2 but with external torque input T. This input makes the system time-varying because in general, $T = T(t)$. The equation for this system is

$$T - mgl\sin\theta - b\dot{\theta} = ml^2\ddot{\theta} \tag{3.8}$$

Using

$$\begin{aligned} x_1 &= \theta \\ x_2 &= \dot{\theta} \end{aligned} \tag{3.9}$$

the state equations become

$$\begin{aligned}\dot{x}_1 &= x_2 \\ \dot{x}_2 &= -\frac{g}{l}\sin x_1 - \frac{b}{ml^2}x_2 + \frac{1}{ml^2}T\end{aligned} \tag{3.10}$$

Putting these state equations in the form of (3.3) gives

$$f(x_1, x_2, T) = \begin{bmatrix} f_1(x_1, x_2, T) \\ f_2(x_1, x_2, T) \end{bmatrix} \tag{3.11}$$

where

$$\begin{aligned}f_1(x_1, x_2, T) &= x_2 \\ f_2(x_1, x_2, T) &= -\frac{g}{l}\sin x_1 - \frac{b}{ml^2}x_2 + \frac{1}{ml^2}T\end{aligned} \tag{3.12}$$

First, is this system piecewise continuous in t? It depends. Although t does not appear explicitly in the system equations, it does appear implicitly through the torque input T. Thus, if T is piecewise continuous, then so is our system. This result comes about because f_2 is a continuous function of T, and continuous functions of continuous functions are also continuous.

Next, is f a Lipschitz function? One approach is to pick an arbitrary pair of points in the plane, $x = (x_1, x_2)$ and $y = (y_1, y_2)$ and then find a constant L (which is independent of x and y) so that (3.5) is satisfied. Instead of this approach, we will rather use the fact that continuously differentiable functions satisfy the Lipschitz condition. This is stated formally as a theorem.[3]

Theorem: *Let $f(t, x)$ be continuous on $[a, b] \times \Re^N$. If $\frac{\partial f}{\partial x}$ exists and is continuous on $[a, b] \times \Re^N$, then f is globally Lipschitz in x on $[a, b] \times \Re^N$ if and only if $\frac{\partial f}{\partial x}$ is uniformly bounded on $[a, b] \times \Re^N$.*

To apply this theorem, we first need $\frac{\partial f}{\partial x}$, the Jacobian of f, which is given by

$$\begin{aligned}\frac{\partial f}{\partial x} &= \begin{bmatrix} \dfrac{\partial f_1}{\partial x_1} & \dfrac{\partial f_1}{\partial x_2} \\ \dfrac{\partial f_2}{\partial x_1} & \dfrac{\partial f_2}{\partial x_2} \end{bmatrix} \\ &= \begin{bmatrix} 0 & 1 \\ -\dfrac{g}{l}\cos x_1 & -\dfrac{b}{ml^2} \end{bmatrix}\end{aligned} \tag{3.13}$$

The Jacobian certainly exists and is continuous. Next is to determine if $\frac{\partial f}{\partial x}$ is uniformly bounded on $[a, b] \times \Re^2$. This condition also holds because none of the entries in the matrix grows without bound (the cosine gives values between -1 and 1 and all the other entries in the matrix are constant). Thus, we can conclude that the system equations for the damped pendulum with torque input are globally Lipschitz. The "globally" modifier means that the condition holds on all of plane $\Re^2$, not just a subset of it.

Finally, does (3.6) hold? To determine this condition, consider any initial condition $x_0 = (x_{10}, x_{20})$.

Then for $t \in [t_0, t_1]$,

$$f(t, x_0) = \begin{bmatrix} x_{20} \\ -\frac{g}{l}\sin x_{10} - \frac{b}{ml^2}x_{20} + \frac{1}{ml^2}T(t) \end{bmatrix} \tag{3.14}$$

The dependence of torque T on time is shown explicitly. Then using the Euclidean distance norm

$$\|f(t, x_0)\| = \sqrt{x_{20}^2 + \left(-\frac{g}{l}\sin x_{10} - \frac{b}{ml^2}x_{20} + \frac{1}{ml^2}T(t)\right)^2} \tag{3.15}$$

If T is bounded on the interval $t \in [t_0, t_1]$, then (3.6) is satisfied.

Because all the conditions of the existence and uniqueness theorem are satisfied with certain restrictions on T, we can conclude that the damped pendulum equations have a unique solution.

Example 2: Next, let us consider a system with a dead zone nonlinearity.

$$\dot{x} = ax + b\phi(x) \tag{3.16}$$

where the nonlinearity is defined by

$$\phi(x) = \begin{cases} x + d, & \text{for} \quad x < -d \\ 0, & \text{for} \quad -d \le x \le d \\ x - d, & \text{for} \quad x > d \end{cases} \tag{3.17}$$

Dead zone is a common nonlinearity that often shows up in physical systems when there is some type of "stickiness." This type of nonlinearity will appear again in the next chapter when discussing nonlinearities. A plot of $f(x)$ versus x with $a > 0$ and $b > 0$ is shown in Figure 3.4.

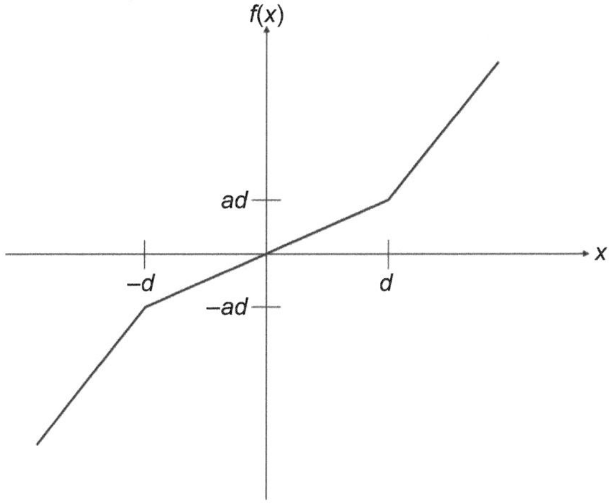

Figure 3.4 Plot of the system exhibiting a dead zone linearity.

Applying the existence and uniqueness theorem, we first note that $f(x)$ is continuous in t. Continuity can be established by the fact that f is a continuous function of x (as seen in Figure 3.4) and x is a continuous (actually differentiable because $\dot{x}$ exists for all t) function of t.

Next is to check the Lipschitz condition, which is done using a geometric interpretation. In one dimension, the norm $\|\bullet\|$ can be interpreted as the absolute value. Then (3.5) becomes

$$|f(x) - f(y)| \leq L|x - y| \tag{3.18}$$

or

$$\frac{|f(x) - f(y)|}{|x - y|} \leq L \tag{3.19}$$

The left side of (3.19) is the slope of the line connecting two points on the function f. The Lipschitz condition then reduces to having a bound on the slope of the line connecting any two points. Choosing the two points x and y in the same region (i.e. $x, y < -d$, $x, y \in [-d, d]$, or $x, y > d$), the slope of the line connecting $f(x)$ and $f(y)$ is the slope of that segment (a or $a + b$). If the points are in different regions, the slope connecting $f(x)$ and $f(y)$ does not exceed $a + b$ as shown in Figure 3.5. No matter where x and y are located, the slope of the line will never be steeper than the steepest segment of f.

Checking the final condition gives

$$\begin{aligned} |f(t, x_0)| &= |ax_0 + b\phi(x_0)| \\ &\leq h \end{aligned} \tag{3.20}$$

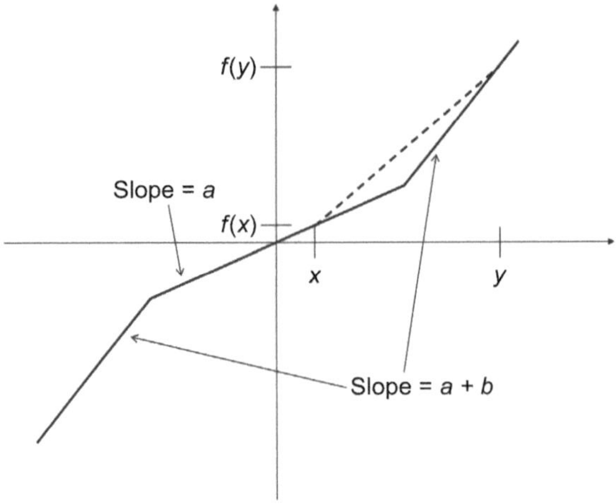

Figure 3.5 A geometric interpretation of the Lipschitz condition as a bound on the slope connecting two points on *f*(x).

This condition holds for each x_0 and all t (because f does not depend explicitly on t). All the conditions of the existence and uniqueness theorem are satisfied, so therefore a unique solution to (3.16) exists.

To summarize: why are we interested in whether a solution exists and whether it is unique? So we are assured that the mathematical model of the system can (1) predict where the system will go next and (2) make the prediction without ambiguity! However, for engineers, this is only the starting point. Equally as important is *knowing* what the solution *is*. In general, finding the solution for the system is difficult. But in the case of linear systems, there is a straightforward procedure for finding them, which we explore next.

3.2.4 Solutions of Linear Systems

To find the solution to an autonomous linear system of the form

$$\dot{x} = Ax \tag{3.21}$$

this system is treated much like a first-order scalar differential equation $\dot{x} = ax$ for which the solution is $x(t) = x_0 e^{at}$ where x_0 is the initial condition for the system. Likewise, the solution to (3.21) is of the form

$$x(t) = e^{At}x_0 \tag{3.22}$$

except now x and x_0 are vectors and the exponential is a matrix.

We consider two cases: A is diagonalizable and A is not diagonalizable. One can check if A is diagonalizable by its eigenvectors. If the set of eigenvectors is linearly independent, then A is diagonalizable.[6]

In the case that **A is diagonalizable**, the exponential can be expanded as

$$e^{At} = e^{\lambda_1 t}G_1 + e^{\lambda_2 t}G_2 + \cdots + e^{\lambda_N t}G_N \tag{3.23}$$

where λ_i denote the eigenvalues and G_i is the ith **spectral projector** given by

$$G_i = \frac{\Pi_{\substack{j=1\\ j\neq i}}^{N}(A - \lambda_j I)}{\Pi_{\substack{j=1\\ j\neq i}}^{N}(\lambda_i - \lambda_j)} \tag{3.24}$$

In the case that **A is not diagonalizable**, the exponential is

$$e^{At} = \sum_{i=1}^{s}\sum_{j=0}^{k_i-1}\frac{t^j e^{\lambda_i t}}{j!}(A - \lambda_i I)^j G_i \tag{3.25}$$

where s is the number of distinct eigenvalues, G_i is a spectral projector (*not* the same as (3.24) as shown in the examples below), and k_i is the index of λ_i, defined as

$$\text{index}(\lambda) = \text{smallest value of } k \text{ such that } \text{rank}(A - \lambda I)^k = \text{rank}(A - \lambda I)^{k+1}$$

Let us illustrate this solution method with two examples, one for each case.

Example 1: Consider the second-order system given by

$$\ddot{x} = -4x \tag{3.26}$$

From studies of differential equations, we know that the solution is sinusoidal, but let's look at it from a systems perspective.

Letting $x_1 = x$ and $x_2 = \dot{x}$, the system can be written as

$$\begin{bmatrix}\dot{x}_1\\ \dot{x}_2\end{bmatrix} = \begin{bmatrix}0 & 1\\ -4 & 0\end{bmatrix}\begin{bmatrix}x_1\\ x_2\end{bmatrix} \tag{3.27}$$

Assume the initial conditions are $x_1(0) = x_{10}$ and $x_2(0) = x_{20}$. The eigenvalues of this system are $\lambda_1 = 2j$ and $\lambda_2 = -2j$ and the eigenvectors are $\begin{bmatrix}\frac{1}{\sqrt{5}} & \frac{2}{\sqrt{5}}j\end{bmatrix}^T$ and $\begin{bmatrix}\frac{1}{\sqrt{5}} & -\frac{2}{\sqrt{5}}j\end{bmatrix}^T$. Because these eigenvectors are linearly

[6] More details can be found in Meyer (2000), an excellent book on matrix analysis.

independent, the system is diagonalizable. Then by (3.22)−(3.24), and Euler's formula, the solution to (3.27) found as

$$G_1 = \frac{\begin{bmatrix} 0 & 1 \\ -4 & 0 \end{bmatrix} - \begin{bmatrix} -2j & 0 \\ 0 & -2j \end{bmatrix}}{2j - (-2j)} = \begin{bmatrix} \frac{1}{2} & -\frac{1}{4}j \\ j & \frac{1}{2} \end{bmatrix} \tag{3.28}$$

$$G_2 = \frac{\begin{bmatrix} 0 & 1 \\ -4 & 0 \end{bmatrix} - \begin{bmatrix} 2j & 0 \\ 0 & 2j \end{bmatrix}}{-2j - 2j} = \begin{bmatrix} \frac{1}{2} & \frac{1}{4}j \\ -j & \frac{1}{2} \end{bmatrix} \tag{3.29}$$

$$\begin{bmatrix} x_1(t) \\ x_2(t) \end{bmatrix} = e^{2jt} \begin{bmatrix} \frac{1}{2} & -\frac{1}{4}j \\ j & \frac{1}{2} \end{bmatrix} \begin{bmatrix} x_{10} \\ x_{20} \end{bmatrix} + e^{-2jt} \begin{bmatrix} \frac{1}{2} & \frac{1}{4}j \\ -j & \frac{1}{2} \end{bmatrix} \begin{bmatrix} x_{10} \\ x_{20} \end{bmatrix}$$

$$= \begin{bmatrix} (\cos(2t) + j\sin(2t))\left(\frac{1}{2}x_{10} - \frac{1}{4}jx_{20}\right) \\ \quad + (\cos(2t) - j\sin(2t))\left(\frac{1}{2}x_{10} + \frac{1}{4}jx_{20}\right) \\ (\cos(2t) + j\sin(2t))\left(jx_{10} + \frac{1}{2}x_{20}\right) \\ \quad + (\cos(2t) - j\sin(2t))\left(-jx_{10} + \frac{1}{2}x_{20}\right) \end{bmatrix}$$

$$= \begin{bmatrix} x_{10}\cos(2t) + \frac{1}{2}x_{20}\sin(2t) \\ -2x_{10}\sin(2t) + x_{20}\cos(2t) \end{bmatrix} \tag{3.30}$$

Example 2: Now consider the system given by

$$\begin{bmatrix} \dot{x}_1 \\ \dot{x}_2 \\ \dot{x}_3 \end{bmatrix} = \begin{bmatrix} -1 & -\frac{1}{2} & -1 \\ -1 & -1 & -1 \\ 1 & \frac{1}{2} & 1 \end{bmatrix} \begin{bmatrix} x_1 \\ x_2 \\ x_3 \end{bmatrix} \tag{3.31}$$

The eigenvalues of this system are $\lambda_1 = -1$ and $\lambda_2 = 0$ (λ_2 has multiplicity 2). and the eigenvectors are $\begin{bmatrix} \frac{1}{\sqrt{6}} & \frac{2}{\sqrt{6}} & -\frac{1}{\sqrt{6}} \end{bmatrix}^T$ and $\begin{bmatrix} \frac{1}{\sqrt{2}} & 0 & -\frac{1}{\sqrt{2}} \end{bmatrix}^T$. Because there are only two unique eigenvectors (instead of three), the system is not diagonalizable. Therefore, to get the solution, (3.25) is needed.

As a preliminary step to applying (3.25), we must determine s, k_i, and G_i. Because there are two distinct eigenvalues, $s = 2$.

For k_i, we determine the index of each eigenvalue by examining the rank of various powers of $(A - \lambda_i I)$.

For $i = 1$,

$$\begin{aligned} \text{rank}\left((A - (-1)I)^1\right) &= 2 \\ \text{rank}\left((A - (-1)I)^2\right) &= 2 \end{aligned}$$

and therefore $k_1 = 1$.

For $i = 2$,

$$\begin{aligned} \text{rank}\left((A - (0)I)^1\right) &= 2 \\ \text{rank}\left((A - (0)I)^2\right) &= 1 \\ \text{rank}\left((A - (0)I)^3\right) &= 1 \end{aligned}$$

and therefore $k_2 = 2$.

Each matrix G_i must have the property that it is the projector onto the nullspace of $(A - \lambda_i I)^{k_i}$ along the range of $(A - \lambda_i I)^{k_i}$. To find the G_1 and G_2, a procedure is applied for building a projector.[7] First define

[7] The complete procedure is found on page 385 of Meyer (2000).

$$\Lambda_1 = (A - (-1)I)^1 = \begin{bmatrix} 0 & -\frac{1}{2} & -1 \\ -1 & 0 & -1 \\ 1 & \frac{1}{2} & 2 \end{bmatrix} \tag{3.32}$$

Then we determine matrices X_1 and Y_1, where the columns of X_1 are a basis of the nullspace of Λ_1 and the columns of Y_1 are a basis of the range of Λ_1.

To find X_1, convert Λ_1 to reduced row echelon form

$$U_1 = \begin{bmatrix} 1 & 0 & 1 \\ 0 & 1 & 2 \\ 0 & 0 & 0 \end{bmatrix} \tag{3.33}$$

and solve for $U_1 v = 0$, which gives

$$\begin{aligned} v_1 + v_3 &= 0 \\ v_2 + 2v_3 &= 0 \end{aligned} \tag{3.34}$$

Choosing v_3 as a free variable, then

$$\begin{bmatrix} v_1 \\ v_2 \\ v_3 \end{bmatrix} = \begin{bmatrix} -1 \\ -2 \\ 1 \end{bmatrix} v_3 \tag{3.35}$$

and

$$X_1 = \begin{bmatrix} -1 \\ -2 \\ 1 \end{bmatrix} \tag{3.36}$$

To find Y_1, we need the basic columns of Λ_1, that is, those that correspond to columns of U_1 containing pivots as in Figure 3.6.

Then

$$Y_1 = \begin{bmatrix} 0 & -\frac{1}{2} \\ -1 & 0 \\ 1 & \frac{1}{2} \end{bmatrix} \tag{3.37}$$

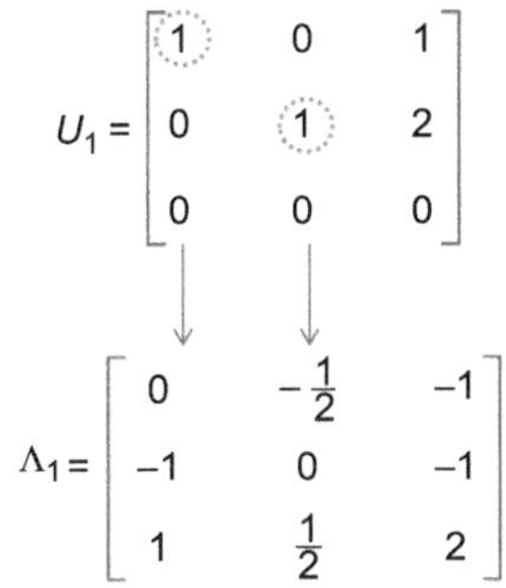

Figure 3.6 The basic columns of Λ_1 correspond to columns of U_1 containing pivots.

and G_1 can be found by

$$G_1 = [X_1|0]\,[X_1|\,Y_1]^{-1}$$

$$= \begin{bmatrix} -1 & 0 & 0 \\ -2 & 0 & 0 \\ 1 & 0 & 0 \end{bmatrix} \begin{bmatrix} -1 & 0 & -\frac{1}{2} \\ -2 & -1 & 0 \\ 1 & 1 & \frac{1}{2} \end{bmatrix}^{-1} \tag{3.38}$$

$$= \begin{bmatrix} \frac{1}{2} & \frac{1}{2} & \frac{1}{2} \\ 1 & 1 & 1 \\ -\frac{1}{2} & -\frac{1}{2} & -\frac{1}{2} \end{bmatrix}$$

Starting with Λ_2 defined as

$$\Lambda_2 = (A-(0)I)^2$$

$$= \begin{bmatrix} \frac{1}{2} & \frac{1}{2} & \frac{1}{2} \\ 1 & 1 & 1 \\ -\frac{1}{2} & -\frac{1}{2} & -\frac{1}{2} \end{bmatrix} \tag{3.39}$$

and following the same procedure yields G_2 as

$$
\begin{aligned}
G_2 &= [X_2|0][X_2|Y_2]^{-1} \\
&= \begin{bmatrix} -1 & -1 & 0 \\ 1 & 0 & 0 \\ 0 & 1 & 0 \end{bmatrix} \begin{bmatrix} -1 & -1 & \frac{1}{2} \\ 1 & 0 & 1 \\ 0 & 1 & -\frac{1}{2} \end{bmatrix}^{-1} \\
&= \begin{bmatrix} \frac{1}{2} & -\frac{1}{2} & -\frac{1}{2} \\ -1 & 0 & -1 \\ \frac{1}{2} & \frac{1}{2} & \frac{3}{2} \end{bmatrix}
\end{aligned}
\tag{3.40}
$$

Now we can proceed with the solution as given in (3.25). Using $s = 2$, $\lambda_1 = -1$, $\lambda_2 = 0$, $k_1 = 1$, $k_2 = 2$, and initial conditions x_{10}, x_{20}, x_{30}, and plugging in gives

$$
\begin{aligned}
\begin{bmatrix} x_1 \\ x_2 \\ x_3 \end{bmatrix} &= \sum_{i=1}^{2} \sum_{j=0}^{k_i-1} \frac{t^j e^{\lambda_i t}}{j!} (A - \lambda_i I)^j G_i \begin{bmatrix} x_{10} \\ x_{20} \\ x_{30} \end{bmatrix} \\
&= \frac{t^0 e^{-t}}{0!}(A - (-1)I)^0 G_1 \begin{bmatrix} x_{10} \\ x_{20} \\ x_{30} \end{bmatrix} + \frac{t^0 e^0}{0!}(A - (0)I)^0 G_2 \begin{bmatrix} x_{10} \\ x_{20} \\ x_{30} \end{bmatrix} \frac{t^1 e^0}{1!}(A - (0)I)^1 G_2 \begin{bmatrix} x_{10} \\ x_{20} \\ x_{30} \end{bmatrix} \\
&= \left[e^{-t} \begin{bmatrix} \frac{1}{2} & \frac{1}{2} & \frac{1}{2} \\ 1 & 1 & 1 \\ -\frac{1}{2} & -\frac{1}{2} & -\frac{1}{2} \end{bmatrix} + \begin{bmatrix} \frac{1}{2} & -\frac{1}{2} & -\frac{1}{2} \\ -1 & 0 & -1 \\ \frac{1}{2} & \frac{1}{2} & \frac{3}{2} \end{bmatrix} + t \begin{bmatrix} -\frac{1}{2} & 0 & -\frac{1}{2} \\ 0 & 0 & 0 \\ \frac{1}{2} & 0 & \frac{1}{2} \end{bmatrix} \right] \begin{bmatrix} x_{10} \\ x_{20} \\ x_{30} \end{bmatrix}
\end{aligned}
\tag{3.41}
$$

Simplifying gives the final solution to (3.31) as

$$\begin{bmatrix} x_1 \\ x_2 \\ x_3 \end{bmatrix} = \begin{bmatrix} \frac{1}{2}e^{-t} + \frac{1}{2} - \frac{1}{2}t & \frac{1}{2}e^{-t} - \frac{1}{2} & \frac{1}{2}e^{-t} - \frac{1}{2} - \frac{1}{2}t \\ e^{t} - 1 & e^{-t} & e^{-t} - 1 \\ -\frac{1}{2}e^{-t} + \frac{1}{2} + \frac{1}{2}t & -\frac{1}{2}e^{-t} + \frac{1}{2} & -\frac{1}{2}e^{-t} + \frac{3}{2} + \frac{1}{2}t \end{bmatrix} \begin{bmatrix} x_{10} \\ x_{20} \\ x_{30} \end{bmatrix} \tag{3.42}$$

This section gave a method and examples of how to find solutions for linear systems, a topic covered rarely in books on dynamical systems. The method is straightforward enough and can be automated using a program such as MATLAB for larger linear systems to take care of the matrix multiplication. As can be expected, finding solutions to nonlinear systems is more difficult. There are many different types of nonlinear systems (indeed, any system that is not linear is nonlinear), and a general procedure for finding an analytical solution is not available. Usually numerical methods are used when working with nonlinear systems.

We now turn our focus to particular types of solutions in dynamical systems.

3.3 EQUILIBRIUM AND NULLCLINES

An equilibrium point is a solution to a dynamical system—a special one. An **equilibrium point** (also known as a **critical point**, **stationary point**, or **fixed point**) is a state of the system where it will stay forever. Mathematically, the equilibrium point is a state of the system x^* that satisfies for discrete-time systems

$$x^*[n+1] = x^*[n] \tag{3.43}$$

and for continuous-time systems

$$\dot{x}^* = f(t, x^*) = 0 \tag{3.44}$$

In other words, when a system gets to its equilibrium point, it doesn't move away from it.

Associated with equilibrium points are curves in the state-space called **nullclines**. For an N-dimensional continuous system with states $x_1, x_2, \ldots, x_N$, the x_i-nullcline is the set of points that satisfies $\dot{x}_i = 0$. The equilibrium points for the system are those points that lie at the intersection of all the nullclines.

Let us explore the concept of equilibrium points in the context of two examples: population dynamics and the double pendulum.

3.3.1 Population Dynamics

Consider the example of a population consisting of predators and their prey. Denote the predator population by x_2 and the prey population by x_1. As a simplified model, the following assumptions are made.

i. If there are no predators, the prey population increases at a rate proportional to its size.
ii. If there are predators, the prey population decreases at a rate proportional to the number of encounters between them.
iii. If there are no prey, the predator population decreases at a rate proportional to its size.
iv. If there are prey, the predator population increases at a rate proportional to the number of encounters between them.
v. The predators only have their prey as a food source.
vi. Effects of overcrowding are ignored.

The equations corresponding to assumptions i to iv are

i. $\dot{x}_1 = ax_1$
ii. $\dot{x}_1 = -bx_1x_2$
iii. $\dot{x}_2 = -cx_2$
iv. $\dot{x}_2 = dx_1x_2$

The constants a, b, c, and d are taken to be positive and depend on how the populations interact. Combining the four equations gives the dynamic equations for the system (known as the Volterra-Lotka system of equations).

$$\dot{x}_1 = (a - bx_2)x_1 \tag{3.45}$$

$$\dot{x}_2 = (-c + dx_1)x_2 \tag{3.46}$$

The x_1-nullclines and x_2-nullclines each have two lines associated with them. From (3.45), the x_1-nullclines are

$$\begin{aligned} x_1 &= 0 \\ x_2 &= \frac{a}{b} \end{aligned} \tag{3.47}$$

From (3.46), the x_2-nullclines are

$$\begin{aligned} x_1 &= \frac{c}{d} \\ x_2 &= 0 \end{aligned} \tag{3.48}$$

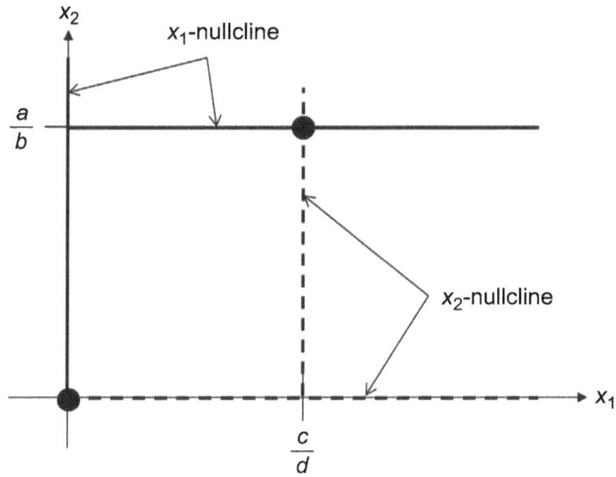

Figure 3.7 The nullclines and equilibrium points of the predator–prey system.

The equilibrium points can be found by setting (3.45) and (3.46) to zero and solving for x_1 and x_2 simultaneously. Thus, the two equilibrium points are

$$\begin{aligned} x_1{}^* &= 0 \text{ and } x_2{}^* = 0 \\ x_1{}^* &= \frac{c}{d} \text{ and } x_2{}^* = \frac{a}{b} \end{aligned} \tag{3.49}$$

Alternatively, the equilibrium points can be found at the intersection of the nullclines as shown in Figure 3.6. In the figure, the x_1-nullclines are shown as the solid lines corresponding to (3.47), and the x_2-nullclines are shown as the dashed lines corresponding to (3.48). The nullclines intersect at two points, (0, 0) and $\left(\frac{c}{d}, \frac{a}{b}\right)$, and these points are the two equilibrium points of the system because both $\dot{x}_1$ and $\dot{x}_2$ are zero.

If the vector fields are plotted on the phase plot in Figure 3.7, there are four distinct behaviors that appear in the four sections created by the nullclines. This behavior is demonstrated in the next MATLAB example.

3.3.1.1 MATLAB Example: Predator–Prey System Phase Plot

Let's now look at the phase plot of the predator–prey system in MATLAB. The code to obtain the plot is very similar to that used in Chapter 2.

```
% predator_prey_simulation.m

% Close all figures and clear all variables
close all
clear all

% Define the model parameters
```

```
a = 4000;
b = 20;
c = 600;
d = 2;

% Define simulation constants
x1max = 500;
x2max = 500;
dx = 40;

% Plot the equilibrium points
hold on
plot(0,0,'ko',c/d,a/b,'ko')

% Plot the nullclines
plot([0,0],[0,x2max],'k','linewidth',2)
plot([0,x1max],[a/b,a/b],'k','linewidth',2)
plot([0,x1max],[0,0],'k--','linewidth',2)
plot([c/d,c/d],[0,x2max],'k--','linewidth',2)

x1 = [0:dx:x1max];
x2 = [0:dx:x2max];
[X1,X2] = meshgrid(x1,x2);

u = (a-b*X2).*X1;
v = (-c+d*X1).*X2;
mag = sqrt(u.^2+v.^2);

% Create the phase plot
quiver(X1,X2,u,v,2)
%quiver(X1,X2,u./mag,v./mag,0.5)
axis([-dx,x1max+dx,-dx,x2max+dx])
xlabel('Prey, x_1')
ylabel('Predators, x_2')
```

After the constants for the simulation are defined, the code first plots the equilibrium points at (0, 0) and $\left(\frac{c}{d}, \frac{a}{b}\right)$ as black circles. Then the nullclines are generated by plotting the two endpoints of the line segment and connecting them with either a solid or dashed black line.

Next a grid is created to locate the positions of each vector field for display on the plot. In this case, the grid is a 500 × 500 area (`x1max` by `x2max`) with points located at intervals of 40 (`dx`). These values were chosen because they gave the best display. The `meshgrid` command creates two 13×13 matrices as follows.

$$X1 = \begin{bmatrix} 0 & 40 & \cdots & 480 \\ 0 & 40 & \cdots & 480 \\ \vdots & \vdots & \ddots & \vdots \\ 0 & 40 & \cdots & 480 \end{bmatrix}$$

$$X2 = \begin{bmatrix} 0 & 0 & \cdots & 0 \\ 40 & 40 & \cdots & 40 \\ \vdots & \vdots & \ddots & \vdots \\ 480 & 480 & \cdots & 480 \end{bmatrix}$$

These two matrices together give the (x_1, x_2) coordinates of each point on the desired grid.

The three lines that follow,

```
u = (a-b*X2).*X1;
v = (-c+d*X1).*X2;
mag = sqrt(u.^2+v.^2);
```

define the vector field with $\dot{x}_1$ denoted by `u` and $\dot{x}_2$ denoted by `v`. Having created `X1` and `X2` with the `meshgrid` command, the calculations for all vector fields on the grid can be accomplished in two lines of code. The last line calculates the magnitude of each vector field and stores it in the variable `mag` so that the vectors of unit length may be plotted.

Last, the phase plot is generated as before using the `quiver` command. Two versions are created. The first, shown in Figure 3.8, is the phase plot with the vector field lengths drawn to scale. The other, shown in Figure 3.9, is the phase plot with each vector field scaled to the same length (done by dividing each vector component by `mag`).

The uniform-length vector fields shown in Figure 3.9 allow us to examine the behavior of the plot in each of the sections created by the nullclines. But first, note the behavior of the vector fields on the nullclines. The x_1-nullcline is plotted as a solid line (as in Figure 3.7). By definition, $\dot{x}_1 = 0$ on the x_1-nullcline, so the vectors coming out of points on the solid lines are vertical. Similarly, on the x_2-nullcline plotted as a dashed line, the $\dot{x}_2$ component of each vector field located on it is zero. Thus, the vectors coming out of points on the dashed lines are horizontal.

The points on the nullclines are the only points in the phase space where the vector fields are vertical or horizontal. Everywhere else, both

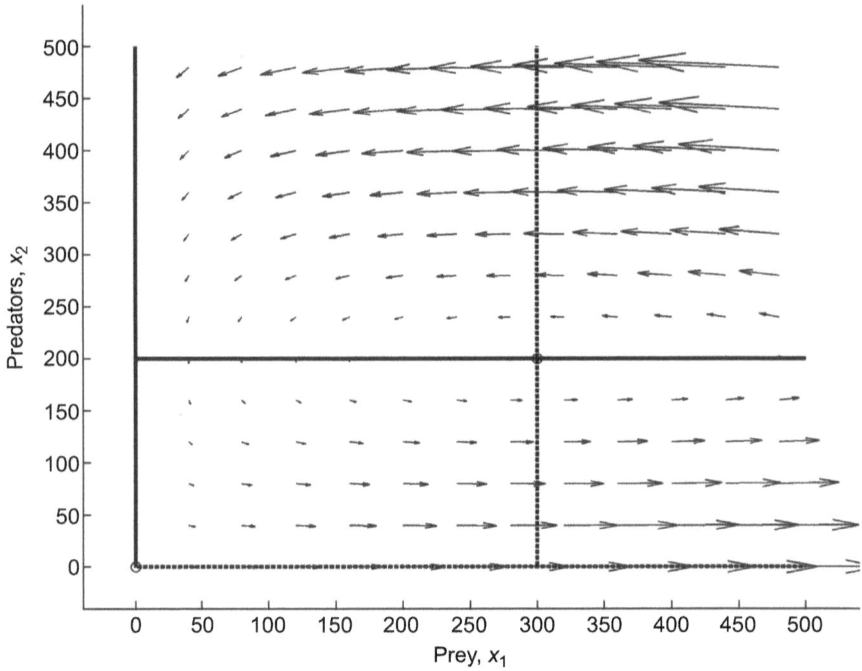

Figure 3.8 The phase plot of the predator–prey system showing the relative lengths of the vector fields.

$\dot{x}_1$ and $\dot{x}_2$ must be nonzero. Therefore they can be identified as pointing northeast, northwest, southeast, or southwest. Using the regions illustrated in Figure 3.10, the behavior of the vector fields is summarized in Table 3.1.

Note that because the system is dealing with populations, only positive x_1 and x_2 are considered. Also in these regions, the exact angle of the vector field isn't specified; only general direction is.

3.3.2 Double Pendulum

A double pendulum is shown in Figure 3.11. Mass m_1 is connected to a fixed point by a massless rod of length l_1. Mass m_2 is connected to m_1 through a massless rod of length l_2. Intuitively, we know that the double pendulum has four configurations in which the segments will remain stationary if placed there carefully and not disturbed. These four configurations are shown in Figure 3.12 and correspond to

a. $\theta_1 = 0, \quad \theta_2 = 0$
b. $\theta_1 = 0, \quad \theta_2 = \pi$
c. $\theta_1 = \pi, \quad \theta_2 = \pi$
d. $\theta_1 = \pi, \quad \theta_2 = 0$

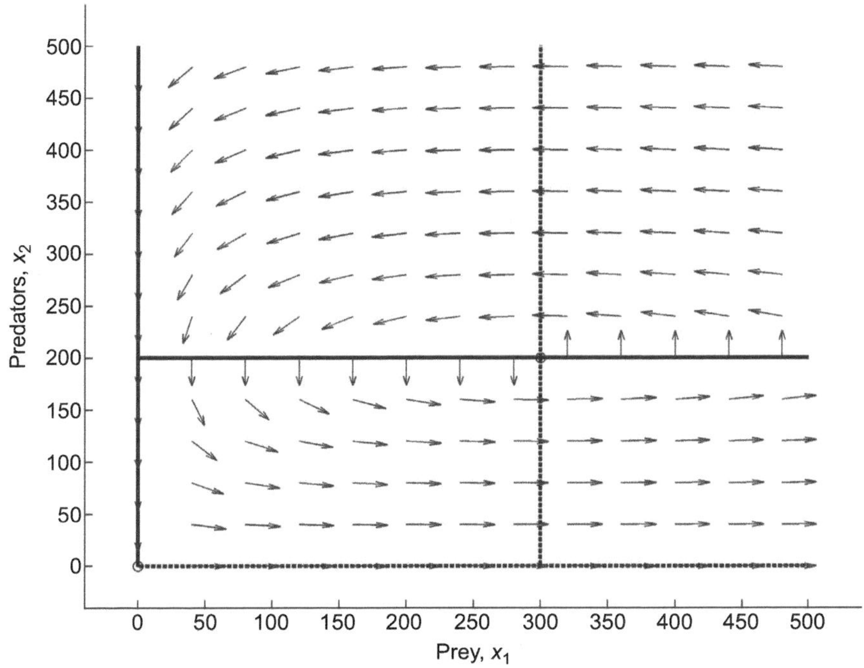

Figure 3.9 The phase plot of the predator–prey system with each vector field scaled to the same length.

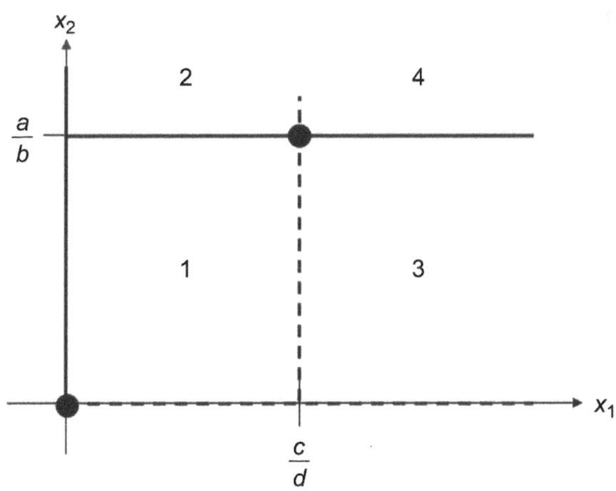

Figure 3.10 The four regions of the phase plot are labeled 1, 2, 3, and 4.

Table 3.1 Summary of Phase Plot Behavior

Region	Location	Vector Field Components	Vector Field Direction
1	$x_1 < \frac{c}{d}$, $x_2 < \frac{a}{b}$	$\dot{x}_1 > 0$, $\dot{x}_2 < 0$	Right, Down } Southeast
2	$x_1 < \frac{c}{d}$, $x_2 > \frac{a}{b}$	$\dot{x}_1 < 0$, $\dot{x}_2 < 0$	Left, Down } Southwest
3	$x_1 > \frac{c}{d}$, $x_2 < \frac{a}{b}$	$\dot{x}_1 > 0$, $\dot{x}_2 > 0$	Right, Up } Northeast
4	$x_1 > \frac{c}{d}$, $x_2 > \frac{a}{b}$	$\dot{x}_1 < 0$, $\dot{x}_2 > 0$	Left, Up } Northwest

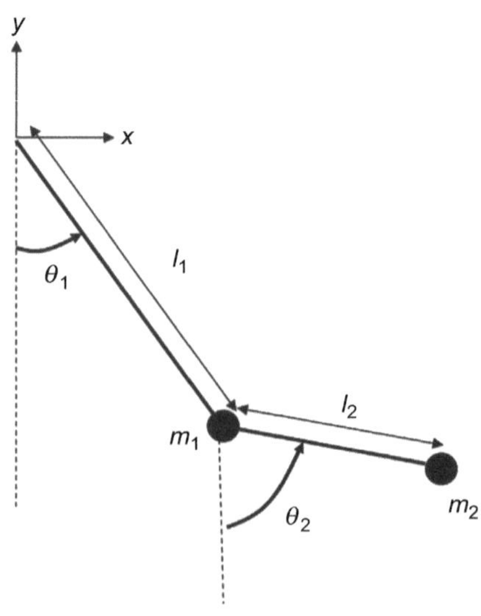

Figure 3.11 The double pendulum.

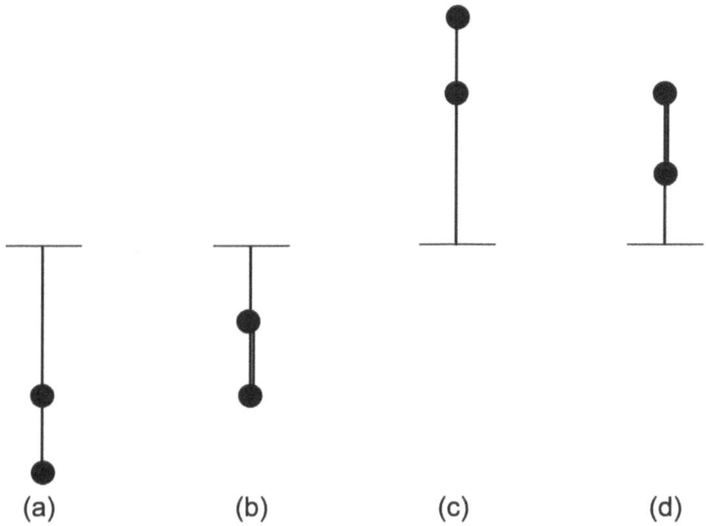

Figure 3.12 The four configurations of equilibrium for the double pendulum.

Although these geometric positions correspond to equilibrium points of the double pendulum, there are, in fact, infinitely many equilibrium points. The reason for this is the circular nature of the pendulum, and adding multiples of 2π gives a different *state* with the same physical position.

Although we have a physical understanding of equilibrium points for the double pendulum, let's explore the mathematics.

To derive the dynamical equations for this system, the Lagrangian technique can be used.[8] The equations are

$$\begin{aligned}(m_1 + m_2)l_1\ddot{\theta}_1 + m_2 l_2 \ddot{\theta}_2 \cos(\theta_1 - \theta_2) + m_2 l_2 \dot{\theta}_2^2 \sin(\theta_1 - \theta_2) \\ + g(m_1 + m_2)\sin\theta_1 = 0\end{aligned} \tag{3.50}$$

$$l_2\ddot{\theta}_2 + l_1\ddot{\theta}_1 \cos(\theta_1 - \theta_2) - l_1\dot{\theta}_1^2 \sin(\theta_1 - \theta_2) + g\sin\theta_2 = 0 \tag{3.51}$$

To determine f for this system, we first define the states to be

$$\begin{aligned} x_1 &= \theta_1 \\ x_2 &= \dot{\theta}_1 \\ x_3 &= \theta_2 \\ x_4 &= \dot{\theta}_4 \end{aligned} \tag{3.52}$$

[8] The derivation of these equations can be found in Weisstein, n.d.

and then put the system equations into the form

$$\begin{aligned} \dot{x}_1 &= f_1(x_1, x_2, x_3, x_4) \\ \dot{x}_2 &= f_2(x_1, x_2, x_3, x_4) \\ \dot{x}_3 &= f_3(x_1, x_2, x_3, x_4) \\ \dot{x}_4 &= f_4(x_1, x_2, x_3, x_4) \end{aligned} \tag{3.53}$$

Notice that the system is time-invariant, so f will not depend explicitly on t. Based on the definition of the states, f_1 and f_3 are simply

$$f_1 = x_2 \tag{3.54}$$

$$f_3 = x_4 \tag{3.55}$$

To obtain f_2, solve (3.51) for $\dot{x}_4$ and plug into (3.50). This gives

$$f_2 = \frac{-m_2\left(l_1 x_2^2 + l_2 x_4^2\right)\sin(x_1 - x_3) - g(m_1 + m_2)\sin x_1 + m_2 g \sin x_3}{l_1(m_1 + m_2 - m_2\cos(x_1 - x_3))} \tag{3.56}$$

Similarly, to obtain f_4, solve (3.50) for $\dot{x}_2$ and plug into (3.51) to get

$$f_4 = \frac{\left(\frac{m_2 l_2}{m_1+m_2} x_4^2 \sin(x_1 - x_3) + g \sin x_1\right)\cos(x_1 - x_3) + l_1 x_2^2 \sin(x_1 - x_3) - g \sin x_3}{l_2\left(1 - \frac{m_2}{m_1+m_2}\cos^2(x_1 - x_3)\right)} \tag{3.57}$$

To determine the equilibrium points of the system, we need to find $(x_1{}^*, x_2{}^*, x_3{}^*, x_4{}^*)$ to satisfy

$$f_1 = f_2 = f_3 = f_4 = 0 \tag{3.58}$$

For f_1 and f_3, it is quite straightforward with

$$\begin{aligned} x_2{}^* &= 0 \\ x_4{}^* &= 0 \end{aligned} \tag{3.59}$$

meaning neither of the pendulum arms can be moving.

For f_2 and f_4, substituting (3.59) yields

$$f_2 = \frac{m_2 g \sin x_3{}^* - g(m_1 + m_2)\sin x_1{}^*}{l_1(m_1 + m_2 - m_2\cos(x_1{}^* - x_3{}^*))} \tag{3.60}$$

$$f_4 = \frac{g \sin x_1{}^* \cos(x_1{}^* - x_3{}^*) - g \sin x_3{}^*}{l_2\left(1 - \frac{m_2}{m_1+m_2}\cos^2(x_1{}^* - x_3{}^*)\right)} \tag{3.61}$$

From (3.60) and (3.61), the criteria for $x_1{}^*$ and $x_3{}^*$ becomes

$$m_2 \sin x_3{}^* - (m_1 + m_2)\sin x_1{}^* = 0 \tag{3.62}$$

$$\sin x_1{}^* \cos(x_1{}^* - x_3{}^*) - \sin x_3{}^* = 0 \tag{3.63}$$

Solving (3.62) and (3.63) for $\frac{\sin x_3{}^*}{\sin x_1{}^*}$ and equating them to each other gives

$$\cos(x_1{}^* - x_3{}^*) = \frac{m_1 + m_2}{m_2} \tag{3.64}$$

which has no solution because $\frac{m_1+m_2}{m_2} > 1 \geq \cos(x_1{}^* - x_3{}^*)$. Therefore, the requirement for equilibrium is

$$\begin{aligned} \sin x_1{}^* &= 0 \\ \sin x_3{}^* &= 0 \end{aligned} \tag{3.65}$$

The solution to (3.65) is

$$\begin{aligned} x_1{}^* &= 0,\ \pm\pi,\ \pm 2\pi,\ \pm 3\pi, \ldots \\ x_3{}^* &= 0,\ \pm\pi,\ \pm 2\pi,\ \pm 3\pi, \ldots \end{aligned} \tag{3.66}$$

which matches the intuitive equilibrium positions shown in Figure 3.12.

Another approach to finding the equilibrium positions is to use nullclines. Because the system is fourth order, there will be four nullclines. Using the system definition in (3.54) to (3.57) and setting $f_1 = f_2 = f_3 = f_4 = 0$ gives x_1-nullcline

$$x_2 = 0 \tag{3.67}$$

x_2-nullcline

$$-m_2\left(l_1 x_2^2 + l_2 x_4^2\right)\sin(x_1 - x_3) - g(m_1 + m_2)\sin x_1 + m_2 g \sin x_3 = 0 \tag{3.68}$$

x_3-nullcline

$$x_4 = 0 \tag{3.69}$$

x_4-nullcline

$$\left(\frac{m_2 l_2}{m_1 + m_2} x_4^2 \sin(x_1 - x_3) + g \sin x_1\right)\cos(x_1 - x_3) + l_1 x_2^2 \sin(x_1 - x_3) - g \sin x_3 = 0 \tag{3.70}$$

Unfortunately, because this is a fourth-order system, the mathematics becomes much more difficult to work through. Furthermore, we aren't able to visualize these nullclines because they are subsets of four-dimensional space. The expressions for the x_1-nullcline and x_3-nullcline are relatively simple. The x_1-nullcline is defined by points in four-dimensional space such that $x_2 = 0$. Similarly, the x_3-nullcline is defined by points in four-dimensional space such that $x_4 = 0$. The equations to get the x_2-nullcline and x_4-nullcline are quite complicated. If we were to work through the mathematics, we would need to solve (3.68) and (3.70) to get a relationship between x_1, x_2, x_3, and x_4 for each equation. These relationships would define a subset of four-dimensional space. The intersection of *all* the subsets would yield the equilibrium points.

3.4 STABILITY

As with many concepts in dynamical systems, we have an intuitive understanding of what stability means. We formalize the understanding with mathematics.

What is stability? Informally, stability means something is somehow behaving properly and predictably and is in control. Stable ground doesn't move. A stable stock market doesn't jump around too much. A stable government provides the necessary services for its people. Stability indicates resistance to change or movement.

This concept applies to the stability of dynamical systems as well. A stable system doesn't behave too much in a crazy or unpredictable manner. It doesn't go out of control or vary greatly. Of course, we are using well-defined terms in an informal way. But it gives an idea of what is meant by stability in the context of dynamical systems.

We will provide formal definitions of stability below (yes, there's more than one), but we must consider the question: To what are we applying the word *stable*? Does one talk about stable systems? Or stable equilibrium points? Or stable responses? Yes, in fact one can discuss all of these. We will describe what each of these ideas mean in turn and then give examples of real-world systems.

As a motivating example, is the damped pendulum system we've used in several prior examples stable? According to the informal notion of stability, it seems that it is indeed. Its motion is, in some sense, controlled and predictable. If an input torque is applied about the pivot, it will eventually rotate in the same direction as that input. If the torque is removed, it doesn't keep rotating forever. Eventually, it comes to rest.

However, what if we investigate the two equilibrium configurations of the pendulum: $\theta = 0$ in Figure 3.13(a) and $\theta = \pi$ in Figure 3.13(b)? These two configurations certainly behave differently. What happens in each case

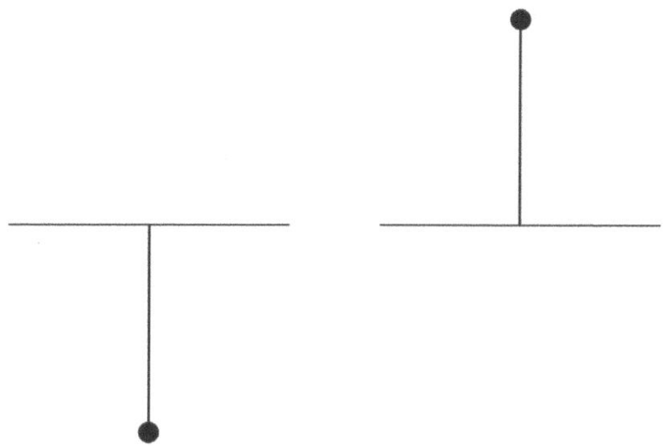

(a) (b)

Figure 3.13 Two equilibrium points of the pendulum. Stable equilibrium (a) and unstable equilibrium (b).

if the pendulum is bumped? For $\theta = 0$, the pendulum will sway back and forth a little bit and eventually come back to $\theta = 0$. By contrast, for $\theta = \pi$, the pendulum will fall down and sway back and forth around $\theta = 0$, eventually settling there. The two different behaviors can be described as (1) returning to the original (stable) equilibrium point and (2) moving away from the original (unstable) equilibrium point and toward the new (stable) one.

With the pendulum example, we see that there are different lenses through which to view stability. Various aspects of a given system can be either stable or unstable (or even marginally stable, a third characterization of stability).

3.4.1 Stable Systems

In the case of linear systems, it is possible to characterize the stability of the entire system. This concept is known as linear stability. Consider the continuous and discrete linear time-invariant systems with no input

$$\dot{x} = Ax \tag{3.71}$$

$$x[n+1] = Ax[n] \tag{3.72}$$

where x is the Nx1 state vector and A is an NxN matrix. It is possible to define three types of stability for a linear system, each with differing levels of strength: **spectral stability**, **linear stability**, and **asymptotic linear stability**.[9]

[9] These definitions are provided in Meiss (2007), with extension to discrete-time systems in Hinrichsen and Pritchard (2005).

3.4.1.1 Spectral Stability

A system described by (3.71) is spectrally stable if all the eigenvalues of A have negative or zero real parts. A system described by (3.72) is spectrally stable if all the eigenvalues of A lie on or within the unit circle.

3.4.1.2 Linear Stability

A system described by (3.71) or (3.72) is linearly stable if all solutions are bounded for all t. That is, for (3.71), if $x(t)$ is a solution, then there exists some M such that

$$\|x(t)\| \leq M \tag{3.73}$$

for all $t \geq 0$.

For (3.72), if $x[n]$ is a solution, then there exists some M such that

$$\|x[n]\| \leq M \tag{3.74}$$

for all $n = 0, 1, 2, \ldots$

3.4.1.3 Asymptotic Linear Stability

A system described by (3.71) is asymptotically linearly stable if all solutions approach zero as $t \rightarrow \infty$. That is, if $x(t)$ is a solution to (3.71), then

$$\lim_{t \rightarrow \infty} \|x(t)\| = 0 \tag{3.75}$$

If $x[n]$ is a solution to (3.72), then

$$\lim_{n \rightarrow \infty} \|x[n]\| = 0 \tag{3.76}$$

These definitions are increasing in strength in the sense that if a system satisfies the stronger condition, it also satisfies the weaker one. For example, if a system is linearly stable, then it is spectrally stable, and if a system is asymptotically linearly stable, then it is both spectrally and linearly stable. Similarly, if a system is not spectrally stable, then it is neither linearly nor asymptotically linearly stable.

To illustrate these concepts, let us consider some simple examples.

Example 1: Consider the first order system

$$\dot{x} = ax \tag{3.77}$$

with initial condition $x(0) = x_0$. The eigenvalue of the system is a, so the system is spectrally stable if $a \leq 0$. The solution to this system is

$$x(t) = x_0 e^{at} \tag{3.78}$$

Table 3.2 Stability of a First Order System

	$a > 0$	$a = 0$	$a < 0$
Solution	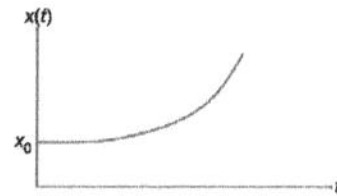		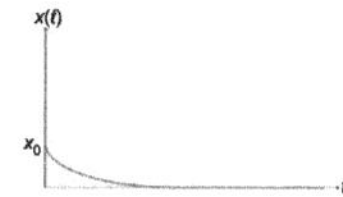
Spectrally stable?	No	Yes	Yes
Linearly stable?	No	Yes	Yes
Asymptotically linearly stable?	No	No	Yes

from which the other types of stability can be determined. For $a = 0$, the solution is constant; thus, it is bounded and linearly stable but not asymptotically linearly stable because it does not approach zero. For $a < 0$, the solution is bounded and approaches zero; thus, it is both linearly and asymptotically linearly stable. A summary of the results is shown in Table 3.2.

Example 2: Consider the second-order system given by

$$\ddot{x} = -4x \tag{3.79}$$

From studies of differential equations, we know that the solution is sinusoidal, but let's look at it from a systems perspective.

Letting $x_1 = x$ and $x_2 = \dot{x}$, the system can be written as

$$\begin{bmatrix} \dot{x}_1 \\ \dot{x}_2 \end{bmatrix} = \begin{bmatrix} 0 & 1 \\ -4 & 0 \end{bmatrix} \begin{bmatrix} x_1 \\ x_2 \end{bmatrix} \tag{3.80}$$

Assume the initial conditions are $x_1(0) = x_{10}$ and $x_2(0) = x_{20}$. The eigenvalues of this system are found to be $\lambda_1 = 2j$ and $\lambda_2 = -2j$. Thus, the system is spectrally stable.

We saw the solution to this system derived in (3.30). It was given as

$$\begin{bmatrix} x_1(t) \\ \\ x_2(t) \end{bmatrix} = \begin{bmatrix} x_{10}\cos(2t) + \frac{1}{2}x_{20}\sin(2t) \\ \\ -2x_{10}\sin(2t) + x_{20}\cos(2t) \end{bmatrix} \tag{3.81}$$

Clearly, the solution is bounded because each involves a sine or cosine that never exceeds a magnitude of 1. Equally as clear is the fact that the solution does not decay to zero over time but oscillates forever. Thus, in this example, the system is linearly stable but not asymptotically stable.

Example 3: In this example, we revisit (3.31), where the system was defined as

$$\begin{bmatrix} \dot{x}_1 \\ \dot{x}_2 \\ \dot{x}_3 \end{bmatrix} = \begin{bmatrix} -1 & -\frac{1}{2} & -1 \\ -1 & -1 & -1 \\ 1 & \frac{1}{2} & 1 \end{bmatrix} \begin{bmatrix} x_1 \\ x_2 \\ x_3 \end{bmatrix} \tag{3.82}$$

From that example, we know the eigenvalues are $\lambda_1 = -1$ and $\lambda_2 = 0$ (with 0 being a repeated eigenvalue). Therefore, the system is spectrally stable.

The solution for this system was found to be

$$\begin{bmatrix} x_1 \\ x_2 \\ x_3 \end{bmatrix} = \begin{bmatrix} \frac{1}{2}e^{-t} + \frac{1}{2} - \frac{1}{2}t & \frac{1}{2}e^{-t} - \frac{1}{2} & \frac{1}{2}e^{-t} - \frac{1}{2} - \frac{1}{2}t \\ e^{t} - 1 & e^{-t} & e^{-t} - 1 \\ -\frac{1}{2}e^{-t} + \frac{1}{2} + \frac{1}{2}t & -\frac{1}{2}e^{-t} + \frac{1}{2} & -\frac{1}{2}e^{-t} + \frac{3}{2} + \frac{1}{2}t \end{bmatrix} \begin{bmatrix} x_{10} \\ x_{20} \\ x_{30} \end{bmatrix} \tag{3.83}$$

This solution is not bounded because the corner entries in the matrix involve a term that is linear in t. These terms grow without bound as t increases. Because there is no upper bound, the system is not linearly stable.

These examples show the different possibilities for stability in linear systems. These definitions of stability cannot be applied to nonlinear systems. One reason is that eigenvalues are not defined for nonlinear systems.[10] Also, nonlinear systems exhibit much more complex behavior. The above definitions, involving long-term characteristics of a solution, are not sufficient to characterize stability in nonlinear systems. The definitions for stability presented next are applicable to general systems and are not specific to only linear systems.

3.4.2 Stable Equilibrium Points

When discussing stability, it is most common to apply the term to equilibrium points, not entire systems as in the previous section. But even when

[10] An attempt at extending eigenvalues and eigenvectors to nonlinear systems can be found in Halas and Moog (2013).

discussing stable equilibrium points, there are three different types to consider: Lyapunov, asymptotic, and exponential. In addition, we will consider what it means for an equilibrium point to be unstable or marginally stable.

3.4.2.1 Lyapunov Stability

An equilibrium point x^* is Lyapunov stable if for every for $\varepsilon > 0$ there exists a $\delta > 0$ such that if $\|x(0) - x^*\| < \delta$, then $\|x(t) - x^*\| < \varepsilon$ for all $t \geq 0$.

Lyapunov stability is implied if an equilibrium point is described simply as "stable."

3.4.2.2 Asymptotic Stability

An equilibrium point x^* is asymptotically stable if it is stable and there exists a $\delta > 0$ such that if $\|x(0) - x^*\| < \delta$, then $\lim_{t \to \infty} \|x(t) - x^*\| = 0$.

3.4.2.3 Exponential Stability

An equilibrium point x^* is exponentially stable if there exists $\alpha > 0$ and $\lambda > 0$ such that $\|x(t) - x^*\| < \alpha e^{-\lambda t}\|x(0) - x^*\|$ for all $t \geq 0$.

As with stability of linear systems, these definitions are increasing in strength. That is, if an equilibrium point is exponentially stable, then it is stable and asymptotically stable. These three definitions can be described qualitatively as well. For stability, a trajectory stays arbitrarily close to the equilibrium point if it doesn't start too far it. For asymptotic stability, a trajectory eventually ends up at the equilibrium point if it doesn't start too far from it. For exponential stability, a trajectory goes to the equilibrium point, and we can specify how fast using an exponential function.

Two additional important terms are important to define: instability and marginal stability.

3.4.2.4 Instability

An equilibrium point x^* is unstable if it is not Lyapunov stable.

3.4.2.5 Marginal Stability

An equilibrium point x^* is marginally stable if it is Lyapunov stable but not asymptotically stable.

All of these definitions become clearer if we think about how system trajectories behave on the phase plane. Figure 3.14 shows what is meant by Lyapunov stability. Given any $\varepsilon > 0$, one can find a $\delta > 0$ so that when a trajectory starts within the ball of radius δ, it stays within the ball of radius ε. The trajectory shown in the figure satisfies this requirement, but for

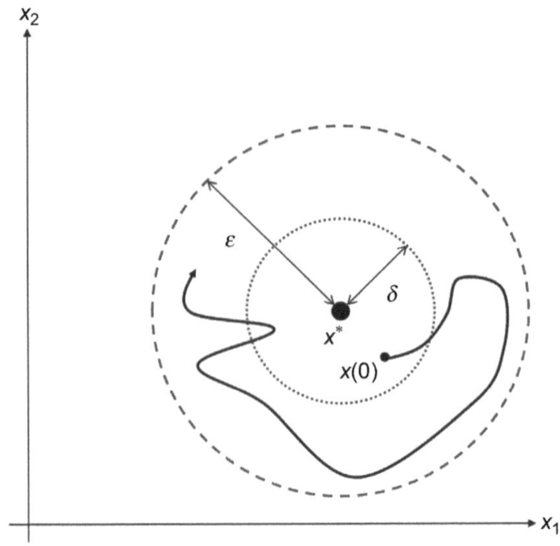

Figure 3.14 Trajectory behavior in a system with a stable equilibrium point.

stability, it *must hold for all possible* trajectories that start within the ball of radius δ. Similarly, Figure 3.15 shows what is meant by asymptotic stability. In this case, the additional requirement is that all trajectories converge to the equilibrium point.

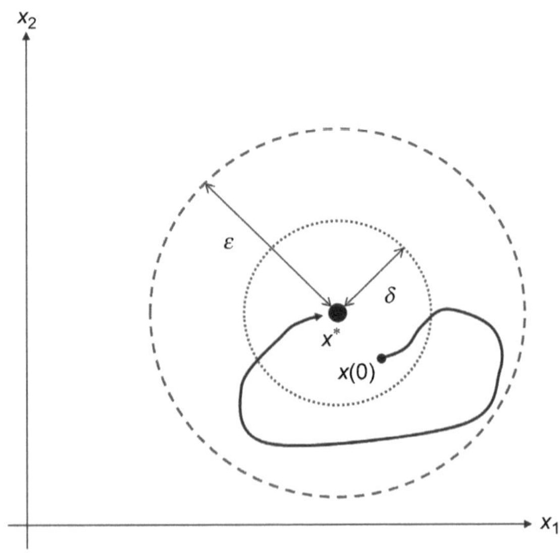

Figure 3.15 Trajectory behavior in a system with an asymptotically stable equilibrium point.

It is important to elaborate on unstable equilibrium points. In linear systems, unstable means the same as "blowing up" in that trajectories move farther and farther away from equilibrium. This behavior is associated in continuous-time systems with eigenvalues that have positive real parts (as shown in the examples following this section). However, nonlinear systems are more complex. They can exhibit instability without trajectories going off to infinity. For example, as shown in Figure 3.16, if a trajectory always moves to a certain distance r from the equilibrium point, no matter where it starts, then it cannot be stable even if it eventually converges to x^*. The reason is that one should be able to choose any $\varepsilon > 0$, and in this case, if $\varepsilon < r$, then we're out of luck choosing δ. Although one can mathematically contrive a system with almost any kind of behavior, this does actually occur in real-world systems in the form of limit cycles. An example of this type of system is discussed in the examples following this section and is also covered in detail in the next chapter.

As a final note in this section, for linear time-invariant systems Lyapunov stability and asymptotic stability come down to checking the eigenvalues of the A matrix. We have the following theorems for stability of linear systems.

Lyapunov Stability for Linear Time-Invariant Systems: *A system* $\dot{x} = Ax$ *is Lyapunov stable if and only if* no *eigenvalues of* A *are in the right half of the complex plane.*

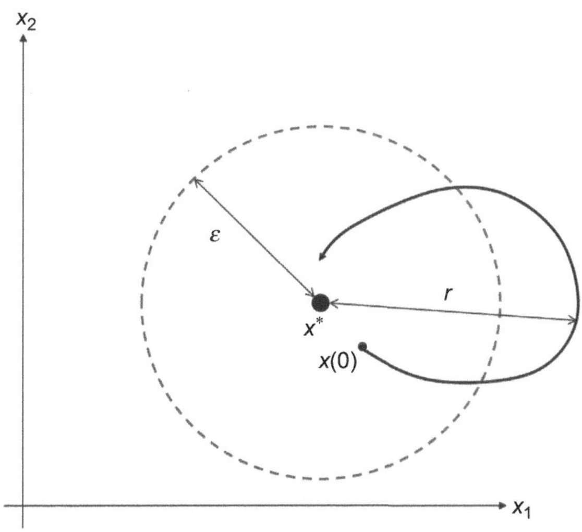

Figure 3.16 An example of unstable behavior even though the trajectories do not "blow up."

Asymptotic Stability for Linear Time-Invariant Systems: *A system $\dot{x} = Ax$ is asymptotically stable if and only if* all *the eigenvalues of A are in the left half of the complex plane.*

The corresponding theorems for discrete-time systems can be obtained by replacing "left (right) half of the complex plane" with "inside (outside) the unit circle."

3.4.3 Stable Responses to an Input

When studying stability of equilibrium points, the effects of external inputs are not considered. Thus, this type of stability is referred to as **internal stability** because it only takes into account how the system behaves on its own. However, in general, systems have inputs, and it is important to know if the system reacts to those inputs in a stable manner. This type of stability is known as **external stability**.

The most common classification of external stability is known as **bounded input, bounded output** (**BIBO**) stability. Another less used classification is **bounded input, bounded state** (**BIBS**) stability. Informally, BIBO and BIBS stability means that the system output or state, respectively, doesn't "blow up" when reasonable inputs are applied to the system.

For the following definitions, assume the system has the continuous-time form with

$$\begin{aligned} \dot{x}(t) &= f(t, x(t), u(t)) \\ y(t) &= h(t, x(t), u(t)) \end{aligned} \tag{3.84}$$

where x is the Nx1 state vector, y is the system output and u is the input.

3.4.3.1 BIBO Stability

The system defined by (3.84) is BIBO stable if for any bounded input $u(t)$ and any initial condition $x(0)$, the output $y(t)$ is also bounded. That is, if there exists an M with $\|u(t)\| \leq M$ for all t, then there exists a constant N_o such that $\|y(t)\| \leq N_o$ for all t.

3.4.3.2 BIBS Stability

The system defined by (3.84) is BIBS stable if for any bounded input $u(t)$ and any initial condition $x(0)$, the state $x(t)$ is also bounded. That is, if there exists an M with $\|u(t)\| \leq M$ for all t, then there exists a constant N_s such that $\|x(t)\| \leq N_s$ for all t.

It is worth mentioning an important difference between linear and nonlinear systems when classifying stability. For linear systems, if one knows its behavior locally (near a certain point in its state-space), those results hold globally throughout the entire state-space. This is a characteristic of linear systems that makes them pleasant to work with. By contrast, nonlinear system global behavior cannot be extrapolated from local behavior. And this characteristic of nonlinear systems also makes them difficult to work with.

Because of this possible change in behavior between different points in the nonlinear system state-space, definitions are often qualified by the term "local" or "global." The definitions given earlier should be considered local when applied to nonlinear systems. For internal stability, it is assumed that the initial condition is near the equilibrium point. If the condition holds for any initial condition, then the equilibrium point exhibits global stability. It is also assumed that the dynamical mappings (such as f and h in (3.84)) are defined on a subset of the state-space about the point of interest. If f and h are defined on the entire state-space, then the BIBO and BIBS stability are global.

Another qualification is made in the case of time-varying systems. The earlier definitions assume the systems are time-invariant. Notice that the starting time is always $t_0 = 0$. In these systems, the behavior depends only on how long the system has been on $(t - t_0)$, not on the absolute time (t). For time-varying systems, the absolute time must be used, and the initial time is given as t_0. When defining stability of equilibrium points, the choice of δ in general depends on t_0 for time-varying systems. If there is no dependence on t_0, then the equilibrium point is **uniformly** stable. For time-invariant systems, stability is always uniform.

3.4.4 Relationship between Types of Stability

With several different types of stability defined, one question is: how are these various notions related? A Venn diagram with these relations for linear systems is shown in Figure 3.17 and for nonlinear systems is shown in Figure 3.18.[11] The relationships for linear systems are more straightforward than for nonlinear systems in that the stronger types are subsets of weaker types. The only exception is the Lyapunov and BIBO stability. One can find examples of systems that exhibit Lyapunov stability that are not BIBO stable and vice versa. With nonlinear systems, the situation is more complicated with less of the structure that has one type of stability implying another. A detailed discussion of these relationships is outside the scope of this book but readers are directed to Khalil (1996), which gives theorems and illustrating examples.

[11] The diagram for linear systems is adopted from Bay (1999).

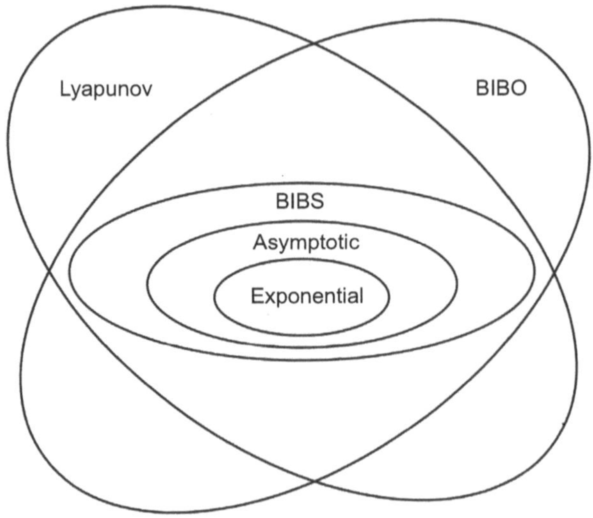

Figure 3.17 The relationship between the different types of stability for linear systems. *BIBO*, bounded input, bounded output; *BIBS*, bounded input, bounded state.

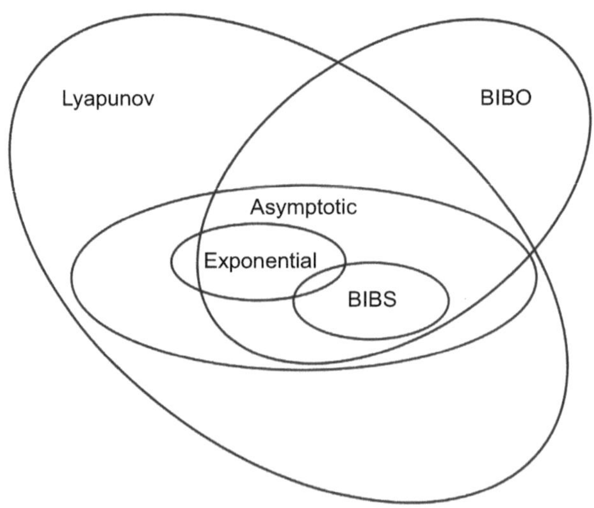

Figure 3.18 The relationship between the different types of stability for nonlinear systems. *BIBO*, bounded input, bounded output; *BIBS*, bounded input, bounded state.

3.4.5 Examples

3.4.5.1 MATLAB Example: Pendulum Stability

Consider again a damped pendulum system

$$\begin{aligned} \dot{x}_1 &= x_2 \\ \dot{x}_2 &= -\frac{g}{l}\sin x_1 - \frac{b}{ml^2}x_2 + \frac{1}{ml^2}T \end{aligned} \tag{3.85}$$

where x_1 is the angle, x_2 is the angular velocity, g is acceleration due to gravity, l is the pendulum length, m is the mass, b is the friction coefficient, and T is the input torque. Let us investigate the different types of stability for this system.

Is the pendulum system spectrally stable, linearly stable, or asymptotically linearly stable? Clearly, it isn't any of these because it's a nonlinear system, and these concepts only apply to linear systems. However, we can investigate the linearized version of (3.85). (Techniques for linearization are covered in detail in the next chapter.) Linearizing about $x = [0, 0]^T$ gives

$$\begin{aligned} \dot{x}_1 &= x_2 \\ \dot{x}_2 &= -\frac{g}{l}x_1 - \frac{b}{ml^2}x_2 + \frac{1}{ml^2}T \end{aligned} \tag{3.86}$$

or in matrix form

$$\begin{bmatrix} \dot{x}_1 \\ \dot{x}_2 \end{bmatrix} = \begin{bmatrix} 0 & 1 \\ -\frac{g}{l} & -\frac{b}{ml^2} \end{bmatrix} \begin{bmatrix} x_1 \\ x_2 \end{bmatrix} + \begin{bmatrix} 0 \\ \frac{1}{ml^2} \end{bmatrix} T \tag{3.87}$$

Setting $T = 0$ because we are checking for an internal type of stability, the eigenvalues are

$$\begin{aligned} \lambda_1 &= \frac{-\frac{b}{ml^2} + \sqrt{\left(\frac{b}{ml^2}\right)^2 - 4\frac{g}{l}}}{2} \\ \lambda_2 &= \frac{-\frac{b}{ml^2} - \sqrt{\left(\frac{b}{ml^2}\right)^2 - 4\frac{g}{l}}}{2} \end{aligned} \tag{3.88}$$

There are three cases to consider: underdamped, critically damped, and overdamped. For the underdamped and critically damped cases, the real part of each eigenvalue is $-\frac{b}{2ml^2}$, and the system is spectrally stable. For the overdamped case, the question is if λ_1 in (3.88) is ever positive. It can be seen that λ_1 is always negative using the fact that

$$\begin{aligned} -4\frac{g}{l} < 0 \quad &\Rightarrow \quad \left(\frac{b}{ml^2}\right)^2 - 4\frac{g}{l} < \left(\frac{b}{ml^2}\right)^2 \\ &\Rightarrow \quad \sqrt{\left(\frac{b}{ml^2}\right)^2 - 4\frac{g}{l}} < \frac{b}{ml^2} \\ &\Rightarrow \quad -\frac{b}{ml^2} + \sqrt{\left(\frac{b}{ml^2}\right)^2 - 4\frac{g}{l}} < 0 \end{aligned} \tag{3.89}$$

Thus, the overdamped pendulum is also spectrally stable.

To check for linear or asymptotic linear stability, we need the solution $x(t)$ to the system in (3.87) when the input T is zero. Following the procedure from Section 3.2.4, the following MATLAB code performs the calculation using the Symbolic Toolbox.

```
% pendulum_solution.m

% Close all figures and clear all variables
close all
clear all

% Define symbolic variables
syms a1 a2 x10 x20 t

% Do the calculation
A = [0, 1; -a1, -a2];
lambda = eig(A);
I = eye(2);
G1 = simplify(A-lambda(2)*I)/(lambda(1)-lambda(2));
G2 = simplify(A-lambda(1)*I)/(lambda(2)-lambda(1));

x = (exp(lambda(1)*t)*G1+exp(lambda(2)*t)*G2)*[x10;x20]
pretty(x)
```

The Symbolic Toolbox allows the user to do calculations with symbols instead of numbers, which is useful here because we haven't specified the parameters of the system and want a general solution. In the code, the first step is to declare the symbolic variables we're using (`a1`, `a2`, `x10`, `x20`, and `t`). Then the `A` matrix is defined and the eigenvalues are found using the `eig` command. The `eye(2)` command defines the 2x2 identity matrix. The next three lines perform the calculation, first by determining `G1` and `G2` and then finding the solution `x` according to (3.22). The `simplify` command is used to algebraically reduce the expression to its simplest form.

The result for `x` is

```
x =

        x20*(exp(-t*(a2/2 - (a2^2 - 4*a1)^(1/2)/2))/(a2^2 - 4*a1)^(1/2)
- exp(-t*(a2/2 + (a2^2 - 4*a1)^(1/2)/2))/(a2^2 - 4*a1)^(1/2)) + x10*
((exp(-t*(a2/2 - (a2^2 - 4*a1)^(1/2)/2))*(a2/2 + (a2^2 - 4*a1)^(1/2)/2))/
(a2^2 - 4*a1)^(1/2) - (exp(-t*(a2/2 + (a2^2 - 4*a1)^(1/2)/2))*(a2/2 -
(a2^2 - 4*a1)^(1/2)/2))/(a2^2 - 4*a1)^(1/2)) - x10*((a1*exp(-t*(a2/2 -
(a2^2 - 4*a1)^(1/2)/2)))/(a2^2 - 4*a1)^ (1/2) - (a1*exp(-t*(a2/2 + (a2^2 -
4*a1)^(1/2)/2)))/(a2^2 - 4*a1)^(1/2)) - x20*((exp(-t*(a2/2 - (a2^2 - 4*a1)
^(1/2)/2))*(a2/2 - (a2^2 - 4*a1)^(1/2)/2))/(a2^2 - 4*a1)^(1/2) - (exp(-
t*(a2/2 + (a2^2 - 4*a1)^(1/2)/2))*(a2/2 + (a2^2 - 4*a1)^(1/2)/2))/(a2^2 -
4*a1)^(1/2))
```

which is a little hard to decipher. Using the `pretty` command makes the output more digestible for human viewing. The final output of the code is

```
/         / exp(-t #1)   exp(-t #2) \        / exp(-t #1) #2   exp(-t #2) #1 \    \
|     x20 | ---------- - ---------- | + x10 | ------------- - ------------- |    |
|         \    #3           #3      /        \     #3              #3       /    |
|                                                                                |
|        / a1 exp(-t #1)   a1 exp(-t #2) \       / exp(-t #1) #1   exp(-t #2) #2 \ |
| - x10 | ------------- - ------------- | - x20 | ------------- - ------------- | |
\       \      #3              #3       /       \      #3              #3      / /

where

                 a2      #3
       #1 ==     --  -   --
                 2       2

                 a2      #3
       #2 ==     --  +   --
                 2       2

                          2
       #3 ==     sqrt(a2   - 4 a1)
```

Written in a more mathematically pleasing form, the solution is

$$x_1(t) = x_{10}\left(\frac{e^{-t\left(\frac{a_2-\sqrt{a_2^2-4a_1}}{2}\right)}\left(a_2+\sqrt{a_2^2-4a_1}\right) - e^{-t\left(\frac{a_2+\sqrt{a_2^2-4a_1}}{2}\right)}\left(a_2-\sqrt{a_2^2-4a_1}\right)}{2\sqrt{a_2^2-4a_1}}\right)$$

$$+ x_{20}\left(\frac{e^{-t\left(\frac{a_2-\sqrt{a_2^2-4a_1}}{2}\right)} - e^{-t\left(\frac{a_2+\sqrt{a_2^2-4a_1}}{2}\right)}}{\sqrt{a_2^2-4a_1}}\right)$$

$$x_2(t) = a_1 x_{10}\left(\frac{e^{-t\left(\frac{a_2-\sqrt{a_2^2-4a_1}}{2}\right)} - e^{-t\left(\frac{a_2+\sqrt{a_2^2-4a_1}}{2}\right)}}{\sqrt{a_2^2-4a_1}}\right)$$

$$+ x_{20}\left(\frac{e^{-t\left(\frac{a_2+\sqrt{a_2^2-4a_1}}{2}\right)}\left(a_2+\sqrt{a_2^2-4a_1}\right) - e^{-t\left(\frac{a_2-\sqrt{a_2^2-4a_1}}{2}\right)}\left(a_2-\sqrt{a_2^2-4a_1}\right)}{2\sqrt{a_2^2-4a_1}}\right) \tag{3.90}$$

Note that $a_1 = \frac{g}{l}$ and $a_2 = \frac{b}{ml^2}$ to simplify the expression. Although (3.90) is a bit long and unwieldy to work with, it is clear upon inspection that $x_1(t)$ and $x_2(t)$ are bounded as a function of the initial condition, and thus the system is linearly stable. Also, $x_1(t)$ and $x_2(t)$ both asymptotically approach zero because of the exponential with the negative power (in all three damping cases, note from (3.89) that $-\left(a_2 - \sqrt{a_2^2 - 4a_1}\right) < 0$). We can conclude that the linearized pendulum is asymptotically linearly stable.

Now let's linearize the system about its upright position $x = [\pi \ 0]^T$. Then the system becomes

$$\begin{bmatrix} \dot{x}_1 \\ \dot{x}_2 \end{bmatrix} = \begin{bmatrix} 0 & 1 \\ \frac{g}{l} & -\frac{b}{ml^2} \end{bmatrix} \begin{bmatrix} x_1 \\ x_2 \end{bmatrix} \tag{3.91}$$

The eigenvalues of this system are

$$\lambda_1 = \frac{-\frac{b}{ml^2} + \sqrt{\left(\frac{b}{ml^2}\right)^2 + 4\frac{g}{l}}}{2}$$
$$\lambda_2 = \frac{-\frac{b}{ml^2} - \sqrt{\left(\frac{b}{ml^2}\right)^2 + 4\frac{g}{l}}}{2} \tag{3.92}$$

These eigenvalues are always real, and λ_1 is always positive because $4\frac{g}{l} > 0$. So the pendulum system linearized about its upright position is not spectrally stable.

Because of the hierarchy of stability, we know that this system cannot be linearly stable. But let us investigate further using the solution to (3.91). The solution takes the same form as (3.90) but with a_1 replaced by $-a_1$

$$x_1(t) = x_{10}\left(\frac{e^{-t\left(\frac{a_2-\sqrt{a_2^2+4a_1}}{2}\right)}\left(a_2+\sqrt{a_2^2+4a_1}\right) - e^{-t\left(\frac{a_2+\sqrt{a_2^2+4a_1}}{2}\right)}\left(a_2-\sqrt{a_2^2+4a_1}\right)}{2\sqrt{a_2^2+4a_1}}\right)$$
$$+ x_{20}\left(\frac{e^{-t\left(\frac{a_2-\sqrt{a_2^2+4a_1}}{2}\right)} - e^{-t\left(\frac{a_2+\sqrt{a_2^2+4a_1}}{2}\right)}}{\sqrt{a_2^2+4a_1}}\right)$$

$$x_2(t) = a_1 x_{10}\left(\frac{e^{-t\left(\frac{a_2-\sqrt{a_2^2+4a_1}}{2}\right)} - e^{-t\left(\frac{a_2+\sqrt{a_2^2+4a_1}}{2}\right)}}{\sqrt{a_2^2+4a_1}}\right)$$
$$+ x_{20}\left(\frac{e^{-t\left(\frac{a_2+\sqrt{a_2^2+4a_1}}{2}\right)}\left(a_2+\sqrt{a_2^2+4a_1}\right) - e^{-t\left(\frac{a_2-\sqrt{a_2^2+4a_1}}{2}\right)}\left(a_2-\sqrt{a_2^2+4a_1}\right)}{2\sqrt{a_2^2+4a_1}}\right) \tag{3.93}$$

With the change in sign, the exponential terms with the exponent involving $a_2 - \sqrt{a_2^2 + 4a_1}$ blow up as t increases because $a_2 - \sqrt{a_2^2 + 4a_1} < 0$.

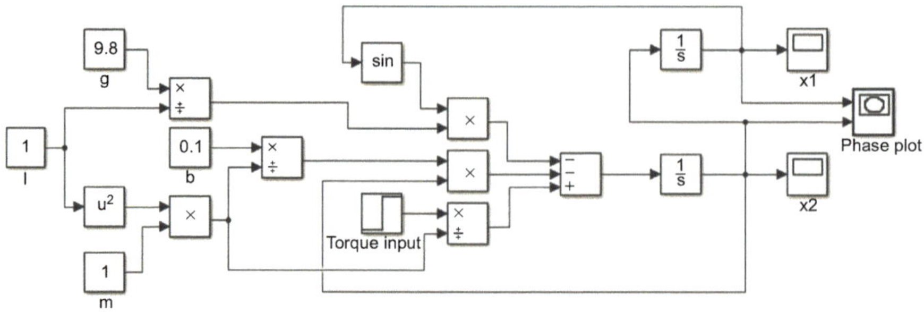

Figure 3.19 Simulink program for the pendulum.

Based on the relationship between stability of a nonlinear system and its linearized counterpart (details are discussed in Section 4.5.2), we can conclude that (0, 0) is a locally asymptotically stable equilibrium point, and (π, 0) is a locally unstable equilibrium point.

The next question to address is whether the pendulum is BIBO stable. To help with this determination, we use Simulink to run the simulation and plot x_1 and x_2 for various values of torque input T. The Simulink program shown in Figure 3.19 implements the equations in (3.86).

The torque input is a step function. The response of the system for a step input of height 5 is shown in Figure 3.20. Both states are converging to a constant value. The angle x_1 settles to a nonzero value, and the angular velocity x_2 goes to zero as expected. After the transient, the pendulum will be held at a constant angle by the torque countering the effect of gravity. This particular bounded input results in bounded states.

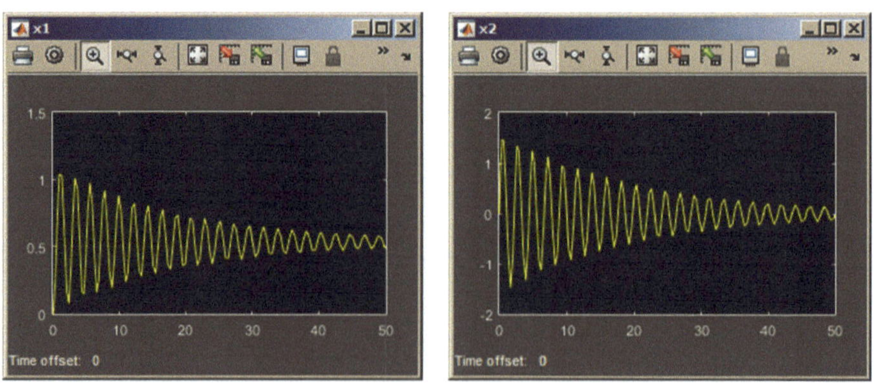

Figure 3.20 The system response with torque input $T = 5$.

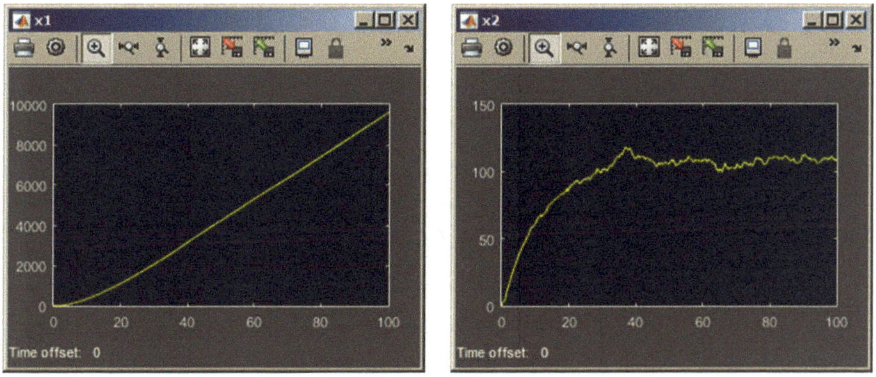

Figure 3.21 The system response with torque input $T = 10$.

Figure 3.21 shows the response to another torque input. This time T is a step input of height 10. In this case, the angle x_1 continues to increase while the angular velocity x_2 levels off. The input torque is large enough so that the pendulum overcomes the gravitational force and continues to swing around its axis.

Is the pendulum BIBO stable? It depends on what you consider its output to be. If the output is the angle x_1, then no. When $T = 10$ (certainly a bounded input), the output grows without bound. It only takes one bounded input resulting in an unbounded output to make the system *not* BIBO stable. However, if the output is taken to be the angular velocity x_2, then yes. The pendulum's rotational speed does not get arbitrarily large.[12] In any case, no matter what your choice of output, the system is not BIBS stable.

3.4.5.2 MATLAB Example: Motor Positioning System

Consider the motor system shown in Figure 3.22. The motor is controlled by a circuit with input $V_s(t)$, and the motor shaft is connected to a load that it must rotate.

The dynamical equations for this system are

$$J_1\ddot{\theta}_1 = -b_1\dot{\theta}_1 - b_2(\dot{\theta}_1 - \dot{\theta}_2) - k(\theta_1 - \theta_2) + K_t i$$

$$J_2\ddot{\theta}_2 = -b_2(\dot{\theta}_2 - \dot{\theta}_1) - k(\theta_2 - \theta_1)$$

$$V_s = Ri + L\frac{di}{dt} + K_e\dot{\theta}_1 \qquad (3.94)$$

[12] This argument does not constitute a proof but rather provides evidence in support of BIBO stability.

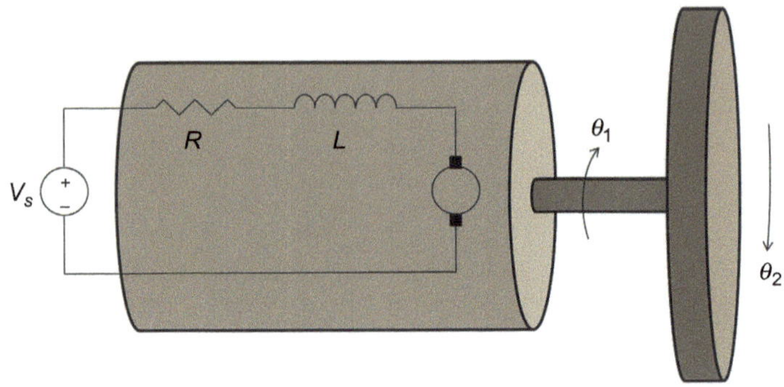

Figure 3.22 Schematic of the motor with a load attached.

where the constants J_1 and J_2 are the inertias of the rotor and load, respectively; b_1 is the friction of the rotor; b_2 is the damping of the shaft; k is the spring constant of the shaft; K_t and K_e are the motor constants; R is the armature resistance; and L is the armature inductance. The system variables are rotor angle θ_1 and angular velocity $\dot{\theta}_1$, load angle θ_2 and angular velocity $\dot{\theta}_2$, and electrical current i. Defining the states to be $x_1 = \theta_1$, $x_2 = \theta_2$, $x_3 = \dot{\theta}_1$, $x_4 = \dot{\theta}_2$, and $x_5 = i$, the input to be $u = V_s$, and the output to be $y = \theta_2$, the state-space form of the system is

$$\begin{bmatrix} \dot{x}_1 \\ \dot{x}_2 \\ \dot{x}_3 \\ \dot{x}_4 \\ \dot{x}_5 \end{bmatrix} = \begin{bmatrix} 0 & 0 & 1 & 0 & 0 \\ 0 & 0 & 0 & 1 & 0 \\ -\dfrac{k}{J_1} & \dfrac{k}{J_1} & -\dfrac{b_1+b_2}{J_1} & \dfrac{b_2}{J_1} & \dfrac{K_t}{J_1} \\ \dfrac{k}{J_2} & -\dfrac{k}{J_2} & \dfrac{b_2}{J_2} & -\dfrac{b_2}{J_2} & 0 \\ 0 & 0 & -\dfrac{K_e}{L} & 0 & -\dfrac{R}{L} \end{bmatrix} \begin{bmatrix} x_1 \\ x_2 \\ x_3 \\ x_4 \\ x_5 \end{bmatrix} + \begin{bmatrix} 0 \\ 0 \\ 0 \\ 0 \\ \dfrac{1}{L} \end{bmatrix} u$$

$$y = \begin{bmatrix} 0 & 1 & 0 & 0 & 0 \end{bmatrix} \begin{bmatrix} x_1 \\ x_2 \\ x_3 \\ x_4 \\ x_5 \end{bmatrix} + [0]u \tag{3.95}$$

Because the system is fifth order, it is difficult to find a general expression for the eigenvalues of A. Instead we use MATLAB to simulate the system for particular values of the parameters.

```
% motor_position.m

% Close all figures and clear all variables
close all
clear all

% Define constants for the system
k = 2000;                % N/m
J1 = 20;                 % kg m^2
J2 = 50;                 % kg m^2
b1 = 25;                 % N s/m
b2 = 30;                 % N s/m
Kt = 5.9e-3;             % Nm/A
Ke = 5.9e-3;             % Nm/A
R = 10;                  % Ohms
L = 0.01;                % Henries

% Define the state space model
A = [0,0,1,0,0;0,0,0,1,0;-k/J1,k/J1,-(b1+b2)/J1,b2/J1,Kt/J1;k/
J2,-k/J2,b2/J2,-b2/J2,0;0,0,-Ke/L,0,-R/L];
B = [0;0;0;0;1/L];
C = [0,1,0,0,0];
D = 0;

lambda = eig(A)
sys_ss = ss(A,B,C,D)
step(sys_ss)
```

The above code returns the eigenvalues

```
lambda =

   1.0e+03 *

  -1.0000 + 0.0000i
  -0.0015 + 0.0117i
  -0.0015 - 0.0117i
   0.0000 + 0.0000i
  -0.0004 + 0.0000i
```

All eigenvalues have negative real parts except for one at the origin. Thus, we can conclude that the system is spectrally stable. We can also conclude that the system is linearly stable but not asymptotically linearly stable. It is linearly stable because there is friction, damping, and resistance in the system to eventually bring x_3, x_4, and x_5 to zero, resulting in x_1 and x_2 achieving a maximum magnitude. It is not asymptotically linearly stable because x_1 and x_2 do not converge to zero. This situation is analogous to (3.77) with $a = 0$.

The equilibrium points of the system can be found from the nullspace of A. That is, all vectors x such that $Ax = 0$. Again, using MATLAB to help with the calculation, the `null` command returns

```
>> null(A)

ans =

  -0.7071
  -0.7071
   0.0000
   0.0000
  -0.0000
```

From this result, we can interpret the equilibrium points of the system to belong to the set in which $x_1 = x_2$ and $x_3 = x_4 = x_5 = 0$. There are infinitely many equilibrium points of the form $[\alpha\ \alpha\ 0\ 0\ 0]^T$, where α is any real number. Looking back at what the system variables are, we can interpret these equilibrium points to be when the motor and load are at the same angle ($x_1 = x_2$), neither the motor nor the load is turning ($x_3 = x_4 = 0$), and there is no current flowing in the circuit ($x_5 = 0$). In this configuration, the system will not change. In any other configuration, the system will change. If $x_1 \neq x_2$, then the spring of the shaft will cause the angles to change. If either $x_3 \neq 0$ or $x_4 \neq 0$, the motor or load is moving but will slow down because of damping. If $x_5 \neq 0$, then there is current in the circuit that will result in torque being applied to the motor shaft.

These equilibrium points are Lyapunov stable but not asymptotically stable. If we choose an equilibrium point and an ε, say $[1\ 1\ 0\ 0\ 0]^T$ and 0.001, then we can always choose initial conditions "close enough" to the equilibrium so that the system settles to within 0.001 of it. In other words, we can ensure that for all $t > 0$,

$$\sqrt{(x_1 - 1)^2 + (x_2 - 1)^2 + x_3^2 + x_4^2 + x_5^2} < 0.001 \tag{3.96}$$

Meaning if we choose carefully enough, we can have the system stop really close to the angle $x_1 = x_2 = 1$. However, we cannot ensure that it will stop exactly at $x_1 = x_2 = 1$ as asymptotic stability requires.

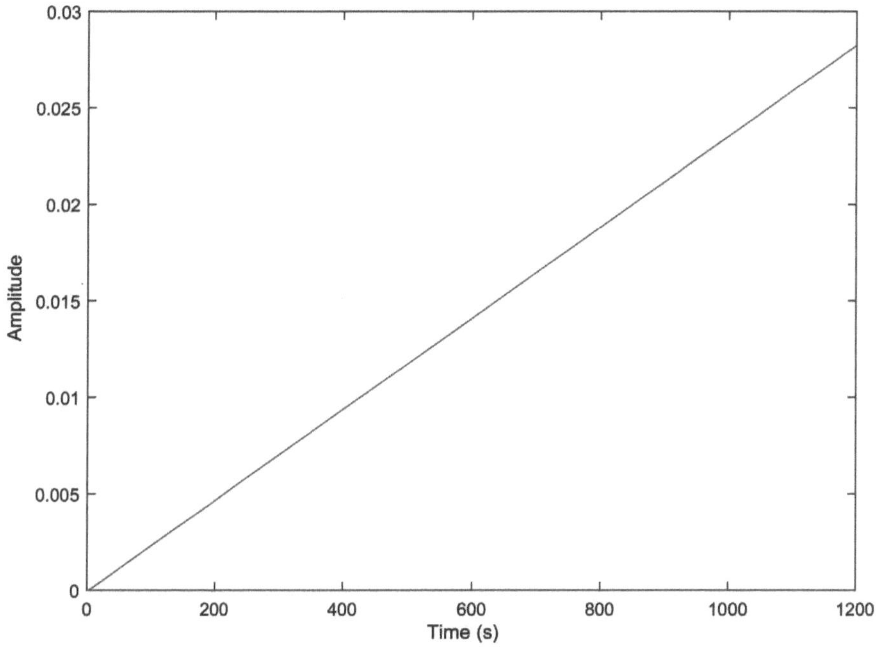

Figure 3.23 The step response of the motor position system with the load angle as the output.

The step response plot is shown in Figure 3.23. From this plot, we can conclude that the system is not BIBO stable. This conclusion makes physical sense. The input to the system is the supply voltage to the motor. When a constant voltage is applied, the motor will turn at a constant speed (in steady state). If the load angle is our output, it will keep increasing without bound.

3.4.5.3 MATLAB Example: Mechanical Belt

Consider the belt system shown in Figure 3.24. A mass m is placed on a belt that is moving with velocity v_0. The mass is connected to a fixed object by means of a spring. The distance of the mass from the object is denoted by x.

The equation of motion for the mass is

$$m\ddot{x} + b\dot{x} + kx + f(\dot{x} - v_0) = 0 \tag{3.97}$$

where k is the coefficient of elasticity; $f(\dot{x} - v_0)$ is the friction of the belt on the mass, which depends their relative velocity; and b represents all other

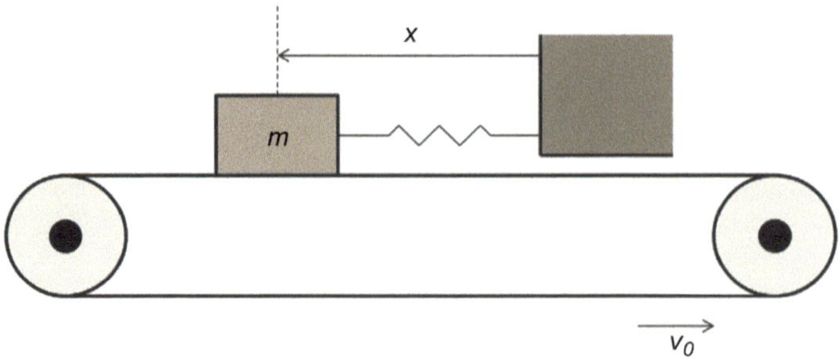

Figure 3.24 A mechanical belt system exhibiting negative friction.

(assumed constant) friction in the system. The function f can be linearized about v_0 (as will be shown in the next chapter), and the equation of motion becomes

$$\ddot{x} + \frac{1}{m}\left(b - \left.\frac{\partial f}{\partial v_0}\right|_{\dot{x}=0}\right)\dot{x} + \frac{k}{m}x = \frac{f(v_0)}{m} \tag{3.98}$$

under the assumption that f is an odd function, meaning that $f(\dot{x} - v_0) = -f(v_0 - \dot{x})$. Defining the offset position y as

$$y = x - \frac{f(v_0)}{k} \tag{3.99}$$

and assuming the friction function takes the form

$$\frac{\partial f}{\partial v_0} = -\alpha y^2 + \beta \tag{3.100}$$

then the system equation becomes

$$\ddot{y} - \frac{1}{m}(\beta - b - \alpha y^2)\dot{y} + \frac{k}{m}y = 0 \tag{3.101}$$

In (3.101), we see a form of the well-known **Van der Pol equation**. Converting to state-space form with $y_1 = y$ and $y_2 = \dot{y}$ gives

$$\begin{aligned} \dot{y}_1 &= y_2 \\ \dot{y}_2 &= \frac{1}{m}(\beta - b - \alpha y_1^2)y_2 - \frac{k}{m}y_1 \end{aligned} \tag{3.102}$$

Setting $\dot{y}_1 = 0$ and $\dot{y}_2 = 0$, the one equilibrium point is found to be $[0, 0]^T$.

The MATLAB code below simulates the system running and generates a phase plot.

```
% mechanical_belt_simulation.m

% Close all figures and clear all variables
close all
clear all

% How long to simulate (in seconds)
t_end = 100;

% Set initial conditions on the system
y1_0 = 4;
y2_0 = 1;

% Solve the system equations
[T Y] = ode45(@mechanical_belt_model,[0 t_end],[y1_0 y2_0]);

% Save the results in a vector
y1 = Y(:,1);
y2 = Y(:,2);

% Plot the results
plot(y1,y2,'b',y1_0,y2_0,'bo')
xlabel('y_1')
ylabel('y_2')
axis equal
axis([-8 8 -8 8])
```

This code for the system model corresponding to (3.102) is shown below.

```
function dy = mechanical_belt_model(t,y)

% Define the model parameters
m = 1;
k = 10;
b = 0.1;
alpha = 1;
beta = 1;

dy = zeros(2,1);
```

```
% Define the dynamical equations
dy(1) = y(2);
dy(2) = (1/m)*(beta-b-alpha*(y(1))^2)*y(2) - (k/m)*y(1);
```

The trajectory for this system with initial condition $[4\ 1]^T$ is shown in Figure 3.25. Although it may look like the trajectory is converging to the origin, it actually does not. It converges to the periodic pattern around the origin. In fact, if the simulation is run for various initial conditions, they all will converge to this same periodic pattern. This pattern is called a **limit cycle**, a phenomenon associated with nonlinear systems discussed further in the next chapter. Figure 3.26 shows the trajectory for an initial condition starting inside the limit cycle and converging to it.

The origin of this system is not Lyapunov stable, and the reason is this: one cannot choose an initial condition to get the trajectory arbitrarily close to the origin. No matter what initial condition is chosen, it converges to

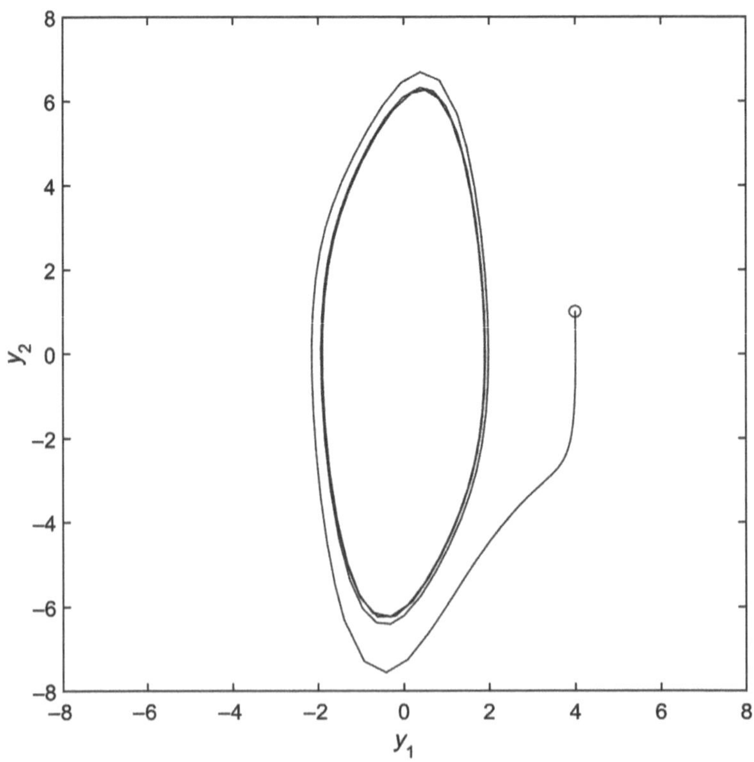

Figure 3.25 Phase plot of the mechanical belt with initial condition $[4\ 1]^T$ exhibiting a limit cycle.

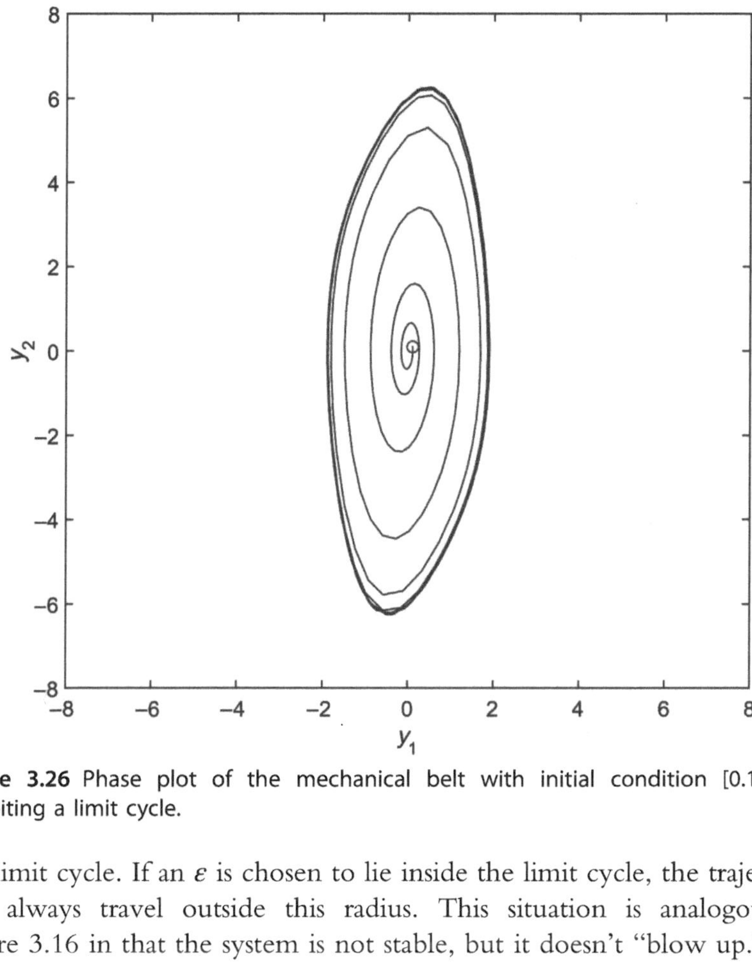

Figure 3.26 Phase plot of the mechanical belt with initial condition $[0.1\ 0.1]^T$ exhibiting a limit cycle.

this limit cycle. If an ε is chosen to lie inside the limit cycle, the trajectory will always travel outside this radius. This situation is analogous to Figure 3.16 in that the system is not stable, but it doesn't "blow up."

3.4.5.4 Example: Automobile Longitudinal Dynamics

Consider the car shown in Figure 3.27. A simple model for the longitudinal dynamics is

$$m\dot{v} = -F_d - mg\sin\theta + F \tag{3.103}$$

where v is the vehicle's longitudinal velocity, F_d is the drag force, m is the vehicle's mass, g is the gravitational acceleration, θ is the road angle of incline, and F is the force imparted from the driven wheels. The drag force depends on how fast the vehicle is traveling and can be modeled as

$$F_d = \frac{1}{2}C_d\rho A v^2\,\mathrm{sgn}(v) \tag{3.104}$$

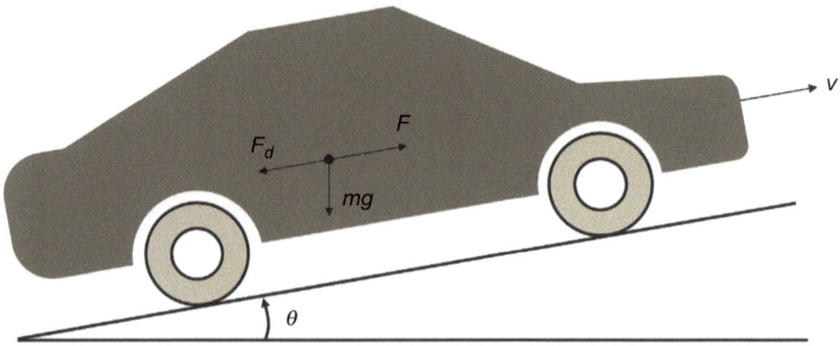

Figure 3.27 The forces acting on the car to determine the longitudinal dynamics.

where C_d is the drag coefficient, ρ is the mass density of air, and A is the cross-sectional area of the vehicle's front. The system is nonlinear because of the sgn(·) function. (This type of nonlinearity is discussed in detail in the next chapter.) So the state equation becomes

$$\dot{v} = -\frac{C_d\rho A}{2m}v^2\text{sgn}(v) - g\sin\theta + \frac{1}{m}F \tag{3.105}$$

If the car is on a flat surface ($\theta = 0$), there is an equilibrium point at $v = 0$. If the car is on any incline ($\theta \neq 0$), then there is an equilibrium point at the velocity for which the force caused by gravity is offset by the drag force. Mathematically, we obtain this result by setting (3.105) to zero and solving for v with $F = 0$.

$$-\frac{C_d\rho A}{2m}v^2\text{sgn}(v) - g\sin\theta = 0 \tag{3.106}$$

If $v > 0$, then (3.106) reduces to

$$v^2 = -\frac{2mg}{C_d\rho A}\sin\theta \tag{3.107}$$

which has no solution for $0 < \theta < 90$ degrees and $v = \sqrt{-\frac{2mg}{C_d\rho A}\sin\theta}$ for -90 degrees $< \theta < 0$.

If $v < 0$, then (3.106) reduces to

$$v^2 = \frac{2mg}{C_d\rho A}\sin\theta \tag{3.108}$$

which has no solution for -90 degrees $< \theta < 0$ and $v = -\sqrt{\frac{2mg}{C_d\rho A}\sin\theta}$ for $0 < \theta < 90$ degrees.

From this calculation, we determined two equilibrium points. When the car is going uphill, there is an equilibrium point at $v = -\sqrt{\frac{2mg}{C_d\rho A}\sin\theta}$. When the car is going downhill, there is an equilibrium point at $v = \sqrt{-\frac{2mg}{C_d\rho A}\sin\theta}$. Note that in the latter case, v is not imaginary because $\sin\theta < 0$.

What can we say about the stability of this equilibrium point? Figure 3.28 shows the trajectories for the uphill case that result when the car starts slightly faster and slightly slower than the equilibrium velocity. Although not a rigorous proof, this demonstrates that the car exhibits asymptotic stability at the equilibrium point, and this matches our intuition about how a car on a hill behaves.

Now to consider BIBO stability, we let F be nonzero. Figure 3.29 shows the response for several values of F. In each case, the vehicle's velocity is bounded. As with the case for no external input, the car settles to a speed for which the all of the forces acting on it (gravitational, drag, and now external) balance, and there is no acceleration. Based on this response, the car's longitudinal velocity exhibits BIBO stability.

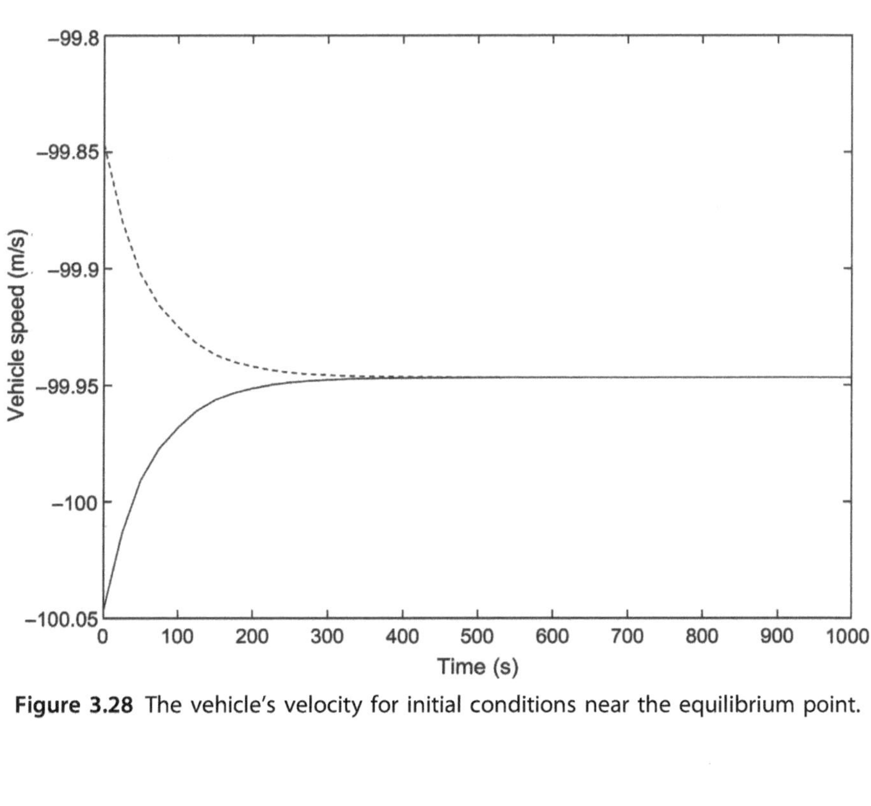

Figure 3.28 The vehicle's velocity for initial conditions near the equilibrium point.

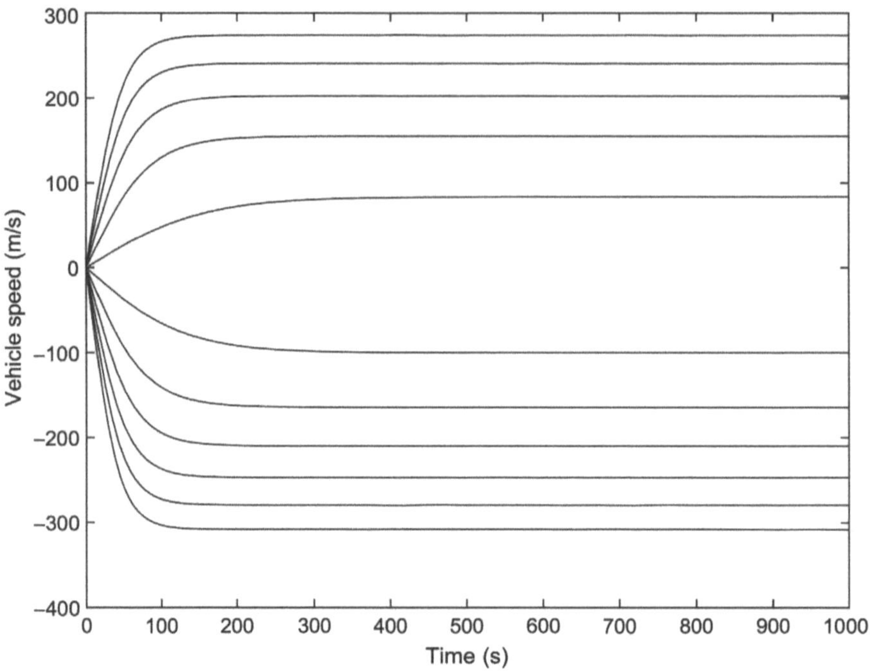

Figure 3.29 The vehicle velocity response for several values of input force.

3.5 LYAPUNOV FUNCTIONS

In the examples given in the previous section, Lyapunov stability was established by applying the definitions and using solutions to the system equations. If the investigation is performed using closed form, analytical solutions, this method will rigorously establish stability of equilibrium points. However, if such a closed form solution is difficult to obtain or does not exist, numerical solutions can be used (as in Section 3.4.5.3). However, numerical solutions cannot rigorously establish stability for they can only give an indication.

There are two other ways besides applying the definition of Lyapunov stability, known as **Lyapunov's indirect (or first) method** and **direct (or second) method**.

Lyapunov's indirect method establishes the local stability of an equilibrium point by checking the behavior of the linearized system about that point. This method is discussed in detail in the next chapter on linearization techniques.

Lyapunov's direct method establishes the local or global stability of an equilibrium point by checking the "energy" of the system through a Lyapunov function. This method is the topic of this section.

The idea behind Lyapunov's direct method is that if the energy in a mechanical system is dissipated, the system will eventually settle down to some equilibrium. Often this concept is introduced by means of the simplest mechanical system: a mass–spring–damper as shown in Figure 3.30. In this system, a mass moves along a frictionless surface and is connected to an immoveable object by a spring and damper.

The equation of motion for this system is

$$m\ddot{x} + b\dot{x} + kx = 0 \tag{3.109}$$

where m is the mass, b is the damping, and k is the spring constant. The energy of the system is the combined kinetic and potential energies given by

$$\begin{aligned} E &= E_{kinetic} + E_{potential} \\ &= \frac{1}{2}m\dot{x}^2 + \int_0^x k\tilde{x}d\tilde{x} \\ &= \frac{1}{2}\left(m\dot{x}^2 + kx^2\right) \end{aligned} \tag{3.110}$$

Taking the time derivative of the energy yields

$$\dot{E} = m\dot{x}\ddot{x} + kx\dot{x} \tag{3.111}$$

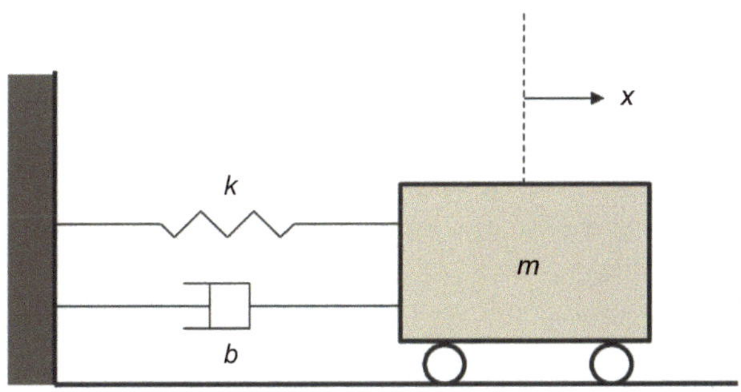

Figure 3.30 The mass–spring–damper system used as motivation for Lyapunov's direct method.

Solving (3.109) for $\ddot{x}$ and substituting into (3.111) gives

$$\begin{aligned} \dot{E} &= m\dot{x}\left(-\frac{b}{m}\dot{x}-\frac{k}{m}x\right)+kx\dot{x} \\ &= -b\dot{x}^2 \end{aligned} \tag{3.112}$$

Because we know the solution $x(t)$ for this linear system, we know that the system is stable if $b = 0$ and asymptotically stable if $b > 0$. In looking at (3.112), we see that $\dot{E} = 0$ if $b = 0$ and $\dot{E} < 0$ if $b > 0$. Furthermore, $E = \dot{E} = 0$ at the equilibrium point $x = 0$, $\dot{x} = 0$. Relating these mathematical results to the system behavior, we can informally conclude (without proof) that for $\dot{E} \leq 0$, the system "stays close" to the equilibrium point, and for $\dot{E} < 0$, the system asymptotically approaches the equilibrium point.

This idea is generalized by defining an energy-like function (the **Lyapunov function**, typically denoted by $V(x)$) with the same characteristics as energy in a mechanical system. There are three important characteristics of the energy function that help in determining stability of the equilibrium point.

1. $V > 0$ is positive away from the equilibrium point.
2. $\dot{V} < 0$ or $\dot{V} \leq 0$ away from the equilibrium point.
3. $V = 0$ at the equilibrium point $x = 0$, $\dot{x} = 0$.

The function V is energy-like because it is positive except for the equilibrium point, where it is zero. Stability of the equilibrium point is established by checking if $\dot{V} \leq 0$ away from equilibrium. The key is that the "energy" is positive and either is bounded or dissipates over time until the system settles.

Here is a formal statement of the Lyapunov theorems for stability and asymptotic stability.

Lyapunov Theorem for Stability: *Let x^* be an equilibrium point of the system $\dot{x} = f(x)$. If there exists a continuous differentiable scalar function $V(x)$ defined in a neighborhood D of x^* such that*

i. $V(x^*) = 0$
ii. $V(x) > 0$ *when* $x \in D$ *and* $x \neq x^*$
iii. $\dot{V}(x) \leq 0$ *when* $x \in D$
then x^* is Lyapunov stable. Additionally, if $V(x)$ satisfies
iv. $\dot{V}(x) < 0$ *when* $x \in D$ *and* $x \neq x^*$,
then x^* is asymptotically stable.

Note the additional condition for asymptotic stability. The time derivative of V is negative, meaning V is always decreasing and approaching zero (it can't go negative because of the second condition). This is the mathematical statement that says "energy" is being dissipated. If the condition is the weaker $\dot{V}(x) \leq 0$, then the "energy" may stop decreasing, but never increases, over time.

There is also a theorem for unstable equilibrium points.

Lyapunov Theorem for Instability: *Let x^* be an equilibrium point of the system $\dot{x} = f(x)$. If there exists a continuous differentiable scalar function $V(x)$ defined in a neighborhood D of x^* such that*

i. $V(x^*) = 0$

ii. $V(x) > 0$ *when* $x \in D$ *and* $x \neq x^*$

iii. $\dot{V}(x) > 0$ *when* $x \in D$

then x^* is unstable.

Because we have the Lyapunov theorem for stability, we only have to find the function $V(x)$ and check some conditions on it, and then we know conclusively if the equilibrium point is stable without having to solve for trajectories in the system.

Isn't this great? Yes and no. On the one hand, it is an advantage to not have to solve the system equations. But on the other hand, there are two significant disadvantages to Lyapunov's theorem.

1. The theorem cannot be used to prove instability in the absence of a Lyapunov function. Notice that the theorem is in the form of an *if—then* statement not an *if-and-only-if* statement. If a function $V(x)$ exists to satisfy the criteria, then one can conclude something about stability. However, if you can't find such a $V(x)$, the equilibrium may or may not be stable. The theorem says nothing about this situation.
2. The theorem gives no direction about how to find such a $V(x)$. The statement starts assuming the Lyapunov function exists. Here again we see the mathematical approach, which talks about the existence of something but ignores the practical side of actually finding it. Unfortunately, there is no general algorithm for finding Lyapunov functions, but experience provides intuition. Also, for certain classes of systems, such as linear time-invariant systems, systematic methods do exist for generating a $V(x)$. But linear systems are *easy* to deal with in the first place. A systematic method for nonlinear systems would be quite useful but unfortunately does not exist. There are some standard ways one can proceed, but they do not guarantee a successful outcome.

3.5.1 Lyapunov Functions for Linear Systems

A typical starting point for constructing Lyapunov functions for linear systems is to make it a quadratic function of the states under the assumption that $x^* = 0$. This assumption ensures that $V(0) = 0$ and $V(x) > 0$ for $x \neq 0$. $V(x)$ is defined as

$$V(x_1, x_2, \ldots, x_N) = \begin{bmatrix} x_1 & x_2 & \cdots & x_N \end{bmatrix} \begin{bmatrix} p_{11} & p_{12} & \cdots & p_{1N} \\ p_{21} & p_{22} & \cdots & p_{2N} \\ \vdots & \vdots & \ddots & \vdots \\ p_{N1} & p_{N2} & \cdots & p_{NN} \end{bmatrix} \begin{bmatrix} x_1 \\ x_2 \\ \vdots \\ x_N \end{bmatrix} \tag{3.113}$$

Or more succinctly,

$$V(x) = x^T P x \tag{3.114}$$

with the restriction that P is a positive definite matrix, which guarantees that $V(x)$ is positive for $x \neq 0$. Recall from matrix theory that a positive definite matrix has all positive eigenvalues, and a positive semidefinite matrix has all nonnegative eigenvalues.

Then taking the derivative

$$\begin{aligned} \dot{V}(x) &= \dot{x}^T P x + x^T P \dot{x} \\ &= (Ax)^T P x + x^T P (Ax) \\ &= x^T A^T P x + x^T P A x \\ &= x^T (A^T P x + x^T P A) x \end{aligned} \tag{3.115}$$

To show stability or asymptotic stability of $x = 0$, one chooses P so that $A^T P x + x^T P A$ is negative semidefinite or negative definite, respectively. Also recall that a matrix M is negative (semi)definite if $-M$ is positive (semi)definite.

Using this formulation, we have a theorem that provides a strong statement about the relationship between P and asymptotic stability.

Theorem for Asymptotic Stability in Linear Time-Invariant Systems: *The origin of the system $\dot{x} = Ax$ is asymptotically stable if and only if given a positive definite matrix Q, the P that solves $A^T P x + x^T P A = -Q$ is also positive definite.*

3.5.1.1 MATLAB Example: Lyapunov Function for a Linearized Pendulum

In this example, we use the MATLAB command `lyap` to find a Lyapunov function for a linearized pendulum modeled by

$$\begin{bmatrix} \dot{x}_1 \\ \dot{x}_2 \end{bmatrix} = \begin{bmatrix} 0 & 1 \\ -\dfrac{g}{l} & -\dfrac{b}{ml^2} \end{bmatrix} \begin{bmatrix} x_1 \\ x_2 \end{bmatrix} \tag{3.116}$$

The code below performs this task.

```
% pendulum_lyap.m

% Close all figures and clear all variables
close all
clear all

% Define constants for the system
g = 9.8;
l = 1;
b = 0.1;
m = 1;

% Define the matrices
A = [0,1;-g/l,-b/(m*l^2)];
Q = eye(size(A));

% Find the Lyapunov function
P = lyap(A,Q)
lambda = eig(P)
```

The `lyap` command solves the equation $A^T Px + x^T PA = -Q$. In using this command for the theorem, the user needs to define Q to be positive semidefinite or positive definite to check for stability or asymptotic stability respectively. The above code returns the following for `P` and `lambda`.

```
P =

     5.5153    -0.5000
    -0.5000    54.0000


lambda =

     5.5102
    54.0052
```

Because the eigenvalues of `P` are positive, it is a positive definite matrix, and we can conclude that the equilibrium point at $x_1 = 0$, $x_2 = 0$ is asymptotically stable.

Setting $b = 0$ and running the code again gives the following result.

```
P =

   1.0e+15 *

   -0.3800   -0.0000
   -0.0000   -3.7239


lambda =

   1.0e+15 *

   -3.7239
   -0.3800
```

In this case, the eigenvalues of P are negative (in fact, they have quite large negative values), making P negative definite. From this, we can conclude that the undamped pendulum is not asymptotically stable, which agrees with our intuition.

Going one step further, setting `b = 0` and `Q = [1,0;0,0]` (which is a positive semidefinite matrix) gives the following result.

```
P =

   1.0e+15 *

   -0.3448   -0.0000
   -0.0000   -3.3791


lambda =

   1.0e+15 *

   -3.3791
   -0.3448
```

This result tells us that the undamped pendulum is Lyapunov stable. The P provided by the `lyap` command solves $A^TPx + x^TPA = -Q$, meaning that $A^TPx + x^TPA$ is negative semidefinite. The existence of such a P is what is important, not the fact that P is negative definite.

For cases in which $x^* \neq 0$, the system can be reworked by redefining the state vector so that $\widehat{x} = x - x^*$, and then the equilibrium is shifted to zero.

3.5.2 Method of Gradients

One method for attempting to construct a Lyapunov function is to use the **method of gradients**.[13] Briefly, the method of gradients assumes the system has the form

$$\begin{bmatrix} \dot{x}_1 \\ \dot{x}_2 \\ \vdots \\ \dot{x}_N \end{bmatrix} = \begin{bmatrix} f_1(x_1, x_2, \ldots, x_N) \\ f_2(x_1, x_2, \ldots, x_N) \\ \vdots \\ f_N(x_1, x_2, \ldots, x_N) \end{bmatrix} \tag{3.117}$$

Then one tries to find a function

$$h(x_1, x_2, \ldots x_N) = \begin{bmatrix} h_1(x_1, x_2, \ldots, x_N) \\ h_2(x_1, x_2, \ldots, x_N) \\ \vdots \\ h_N(x_1, x_2, \ldots, x_N) \end{bmatrix} \tag{3.118}$$

that satisfies three conditions:

i. h is the gradient of the Lyapunov function V, i.e. $h_i = \frac{\partial V}{\partial x_i}$ for $i = 1 \ldots N$

ii. The Jacobian of h is symmetric, i.e. $\frac{\partial h_i}{\partial x_j} = \frac{\partial h_j}{\partial x_i}$ for all $i, j = 1 \ldots N$

iii. $\dot{V}(x) = \frac{\partial V}{\partial x}\dot{x} = h^T f < 0$

Then the functions $h_1, \ldots, h_N$ are integrated to obtain V. The next example shows this method applied to the pendulum system in detail.

3.5.2.1 Example: Lyapunov Function for the Pendulum

Let's apply the method of gradients to the pendulum example with the following dynamical equations.

$$\begin{aligned} \dot{x}_1 &= f_1(x_1, x_2) = x_2 \\ \dot{x}_2 &= f_2(x_1, x_2) = -\frac{g}{l}\sin x_1 - \frac{b}{ml^2}x_2 \end{aligned} \tag{3.119}$$

Now taking the third condition from above results in

$$h_1 x_2 + h_2\left(-\frac{g}{l}\sin x_1 - \frac{b}{ml^2}x_2\right) < 0 \tag{3.120}$$

[13] Full details with proof are provided in Khalil (1996).

Our job is to find some h_1 and h_2 so that (3.120) is satisfied for x_1 and x_2 near the equilibrium point $[0, 0]^T$. This is the point where some creativity will help. Rearranging (3.120) gives

$$h_1 x_2 - h_2 \frac{g}{l} \sin x_1 < h_2 \frac{b}{ml^2} x_2 \tag{3.121}$$

One approach is to try to make the left side of (3.121) to be zero and the right side to be positive, keeping in mind the restriction $\frac{\partial h_1}{\partial x_2} = \frac{\partial h_2}{\partial x_1}$. The choice of

$$\begin{aligned} h_1 &= \alpha_1 x_1 + \beta x_2 + \gamma \frac{g}{l} \sin x_1 \\ h_2 &= \beta x_1 + \alpha_2 x_2 \end{aligned} \tag{3.122}$$

satisfies all these criteria for appropriate choices of α_1, α_2, β, and γ. Plugging (3.122) into (3.120) gives

$$\left(\alpha_1 x_1 + \beta x_2 + \gamma \frac{g}{l} \sin x_1\right) x_2 + (\beta x_1 + \alpha_2 x_2)\left(-\frac{g}{l} \sin x_1 - \frac{b}{ml^2} x_2\right) < 0 \tag{3.123}$$

And simplifying gives

$$\begin{aligned} &\left(\alpha_1 - \beta \frac{b}{ml^2}\right) x_1 x_2 + \left(\beta - \alpha_2 \frac{b}{ml^2}\right) x_2^2 + (\gamma - \alpha_2) \frac{g}{l} x_2 \sin x_1 \\ &- \beta \frac{g}{l} x_1 \sin x_1 < 0 \end{aligned} \tag{3.124}$$

To satisfy (3.124), the following criteria are needed for the constants in h_1 and h_2:

i. $\alpha_1 = \beta \frac{b}{ml^2}$

ii. $\beta < \alpha_2 \frac{b}{ml^2}$

iii. $\gamma = \alpha_2$

iv. $\beta > 0$

The next step is to use h_1 and h_2 to construct V, which can be accomplished through integration.

$$
\begin{aligned}
V(x_1, x_2) &= \int_{x_1^*}^{x_1} h_1(\tilde{x}_1, 0) d\tilde{x}_1 + \int_{x_2^*}^{x_2} h_2(x_1, \tilde{x}_2) d\tilde{x}_2 \\
&= \int_0^{x_1} \left(\alpha_1 \tilde{x}_1 + \gamma \frac{g}{l} \sin \tilde{x}_1\right) d\tilde{x}_1 + \int_0^{x_2} (\beta x_1 + \alpha_2 \tilde{x}_2) d\tilde{x}_2 \\
&= \frac{1}{2} \alpha_1 x_1^2 + \gamma \frac{g}{l} (1 - \cos x_1) + \beta x_1 x_2 + \frac{1}{2} \alpha_2 x_2^2
\end{aligned}
\tag{3.125}
$$

In more succinct form, V can be written as

$$
\begin{aligned}
V(x_1, x_2) &= \frac{1}{2} [x_1 \quad x_2] \begin{bmatrix} \alpha_1 & \beta \\ \beta & \alpha_2 \end{bmatrix} \begin{bmatrix} x_1 \\ x_2 \end{bmatrix} + \gamma \frac{g}{l} (1 - \cos x_1) \\
&= \frac{1}{2} x^T P x + \gamma \frac{g}{l} (1 - \cos x_1)
\end{aligned}
\tag{3.126}
$$

Because V needs to be positive for $x \neq 0$, the two additional conditions on the constants derived from (3.126) are

v. P is a positive definite matrix

vi. $\gamma > 0$

To ensure that P is positive definite, its eigenvalues must be positive. Because the eigenvalues λ_1 and λ_2 are given by

$$
\lambda_1, \lambda_2 = \frac{1}{2} \left(\alpha_1 + \alpha_2 \pm \sqrt{(\alpha_1 + \alpha_2)^2 - 4(\alpha_1 \alpha_2 - \beta^2)} \right) \tag{3.127}
$$

then condition iv can be expressed as

$$
\alpha_1 + \alpha_2 > \sqrt{(\alpha_1 + \alpha_2)^2 - 4(\alpha_1 \alpha_2 - \beta^2)} \tag{3.128}
$$

and simplifying gives an alternative form of the condition above.

v. $\alpha_1 \alpha_2 - \beta^2 > 0$

To satisfy conditions i to vi, the following values can be chosen.

$$\begin{aligned} \alpha_1 &= \frac{1}{2}\left(\frac{b}{ml^2}\right)^2 \\ \alpha_2 &= 1 \\ \beta &= \frac{1}{2}\frac{b}{ml^2} \\ \gamma &= 1 \end{aligned} \tag{3.129}$$

Then by construction, the Lyapunov function and its time derivative are

$$\begin{aligned} V(x_1, x_2) &= \frac{1}{4}\left(\frac{b}{ml^2}\right)^2 x_1^2 + \frac{1}{2}\frac{b}{ml^2}x_1x_2 + \frac{1}{2}x_2^2 + \frac{g}{l}(1 - \cos x_1) \\ \dot{V}(x_1, x_2) &= -\frac{1}{2}\frac{b}{ml^2}\left(\frac{g}{l}x_1 \sin x_1 + x_2^2\right) \end{aligned} \tag{3.130}$$

This function satisfies the conditions of Lyapunov's stability theorem for asymptotic stability. Therefore, we can conclude that the pendulum's equilibrium point at $x^* = 0$ is locally asymptotically stable.

It is worth noting some characteristics of this conclusion. First is that we can only make a statement about local stability because the conditions $V(x_1, x_2) > 0$ and $\dot{V}(x_1, x_2) < 0$ don't apply for all $x \neq 0$. For example, if $x_1 = \pi$ and $x_2 = 0$, then $\dot{V}(x_1, x_2) = 0$. However, $[\pi \ 0]^T$ is another equilibrium point of the system (upright pendulum), and this point is not in the neighborhood of $[0 \ 0]^T$. Because this nonlinear system has multiple equilibrium points, the stability conclusion will only apply locally.

To emphasize again the meaning of the theorem, (3.130) is not the only possible Lyapunov function, and there may be others. The theorem only requires finding one V that satisfies the conditions. When such a V is found, the work is done. If such a V is not found, no conclusion can be made about stability.

This concludes our discussion of characteristics of dynamical systems that can apply to general systems described by differential or difference equations. In particular, we examined in depth: existence and uniqueness of solutions, equilibrium points, and stability. While examining these characteristics, we saw how they were used in applications such as pendulums, vehicular longitudinal dynamics, motor positioning, mechanical belts, and predator–prey systems. In the next chapter, we turn our attention from these general system characteristics to nonlinear system characteristics. We will focus on some concepts that are seen only in nonlinear systems such as limit cycles, bifurcation, chaos, and linearization. We will investigate these concepts in the context of jet engine control, population dynamics, mechanical belts, robotic control, direct current (DC) motors, and pendulum systems.

REFERENCES

Bay, J. (1999). *Fundamentals of Linear State Space Systems*. Boston: WCB/McGraw-Hill.

Halas, M., & Moog, C. H. (2013, September 4–6). *Definition of eigenvalues for a nonlinear system*, 9th IFAC Symposium on Nonlinear Control Systems.

Hinrichsen, D., & Pritchard, A. J. (2005). *Mathematical Systems Theory I: Modeling, State Space Analysis, Stability, and Robustness*. Berlin: Springer-Verlag.

Khalil, H. (1996). *Nonlinear Systems* (2nd ed.). Upper Saddle River, NJ: Prentice-Hall.

Meiss, J. (2007). *Differential Dynamical Systems*. Philadelphia: Society for Industrial and Applied Mathematics.

Meyer, C. (2000). *Matrix Analysis and Applied Linear Algebra*. Philadelphia: Society for Industrial and Applied Mathematics.

Miller, R., & Michel, A. N. (1982). *Ordinary Differential Equations*. New York: Academic Press, Inc.

Schuh, M. (n.d.) Mike Schuh's collection of engineer-mathematician-physicist jokes. Retrieved from http://www.farmdale.com/emp-jokes.shtml.

Weisstein, E. (n.d.). *Double Pendulum*. Retrieved from Eric Weisstein's World of Physics: http://scienceworld.wolfram.com/physics/DoublePendulum.html.

CHAPTER 4

Characteristics of Nonlinear Systems

4.1 TYPES OF NONLINEAR SYSTEMS

In the previous chapter, there was quite a bit of discussion about characteristics of general dynamical systems such as existence and uniqueness of solutions, equilibrium points, and stability. Although some specific results only applied to linear systems, such as certain theorems or solution methods, the concepts themselves are applicable to all systems. In this chapter, linear systems are abandoned to discuss nonlinear systems, only to be picked up again toward the end when discussing linearization.

As mentioned previously, a **nonlinear system** is one that is not linear. In other words, it is a system whose dynamical equations cannot be put in the form

$$\begin{aligned} \dot{x}(t) &= Ax(t) + Bu(t) \\ y(t) &= Cx(t) + Du(t) \end{aligned} \tag{4.1}$$

for continuous-time systems or

$$\begin{aligned} x[n+1] &= Ax[n] + Bu[n] \\ y[n] &= Cx[n] + Du[n] \end{aligned} \tag{4.2}$$

for discrete-time systems as described in Section 2.4.

Nonlinear systems are a *very* broad class of systems because they are defined by what they are not. And because it is such a broad class, general results are hard to come by. Researchers tend to focus on certain types of nonlinearities and generate results for subclasses of systems, those that can be transformed into a certain structure. An example of one type is systems that can be put into chained form (as in Section 2.4.4.6) for which there has been much research in designing controllers. As we begin discussion of nonlinear systems, we will first narrow the focus to a few types of nonlinearities that appear often in real-world systems—elements in systems that exhibit nonlinear characteristics. These elements, which may be found in the system block diagram, can be represented by their

A Practical Approach to Dynamical Systems for Engineers
ISBN 978-0-08-100202-5
http://dx.doi.org/10.1016/B978-0-08-100202-5.00004-8

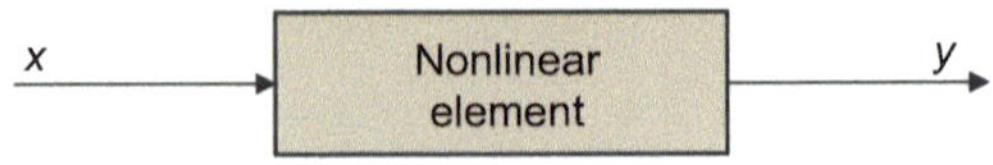

Figure 4.1 Block diagram representation of a nonlinear element.

input—output relation as in Figure 4.1. We will assume the input signal to the block is denoted by x and the output by y.

4.1.1 Relay

The **relay** is represented mathematically by the **sign** or **signum** function. The input—output relationship is expressed as $y = \text{sgn}(x)$, where $\text{sgn}(x)$ can be defined in multiple, but equivalent, ways.

$$\begin{aligned} \text{sgn}(x) &= \begin{cases} 1, & \text{if } x > 0 \\ 0, & \text{if } x = 0 \\ -1, & \text{if } x < 0 \end{cases} \\ &= \begin{cases} \dfrac{x}{|x|}, & \text{if } x \neq 0 \\ 0, & \text{if } x = 0 \end{cases} \end{aligned} \tag{4.3}$$

Figure 4.2 shows y as a function of x, and Figure 4.3 shows the output of the block as a function of time when the input is a sinusoid. This is a simple element that captures the sign of the input. It can be thought of as 1-bit analog-to-digital conversion. It is also the model for Coulomb friction in which the relative velocity of the two surfaces is the input and the resulting force is the output acting in the direction opposite the motion.

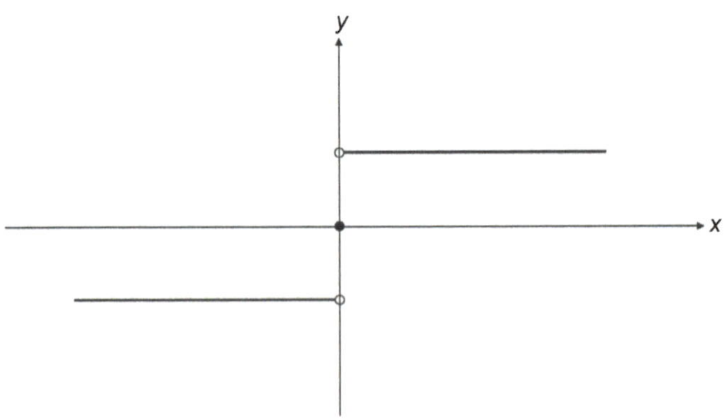

Figure 4.2 The input—output characteristic of a relay or Coulomb friction.

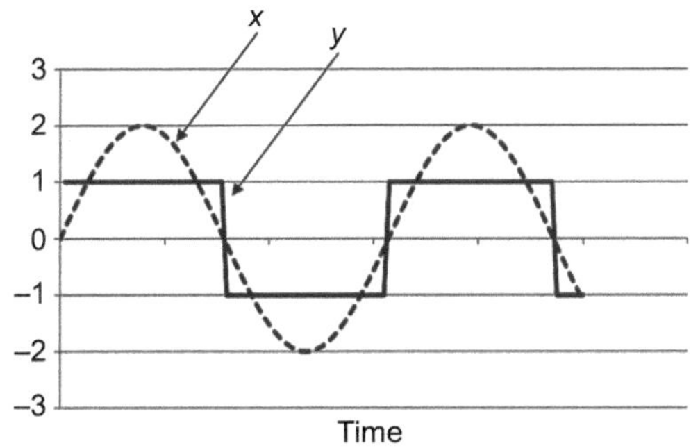

Figure 4.3 The output of the relay when the input is a sinusoid.

This type of nonlinearity can appear in control systems in which a relay (electromechanical switch) is the controller, such as in a thermostat or pump. But we also saw this function in Section 3.4.5.4, the longitudinal dynamics of a vehicle.

4.1.2 Saturation

The characteristic of **saturation** is mathematically expressed as

$$y = \begin{cases} +K & x > x_s \\ ax & -x_s \leq x \leq x_s \\ -K & x < x_s \end{cases} \tag{4.4}$$

An element with saturation nonlinearity has a linear region within input limits. When the input exceeds that limit, the output becomes constant. Figure 4.4 shows y as a function of x, and it is clear that the slope of the function is $a = \frac{K}{x_s}$. The figure also shows the typical behavior in addition to

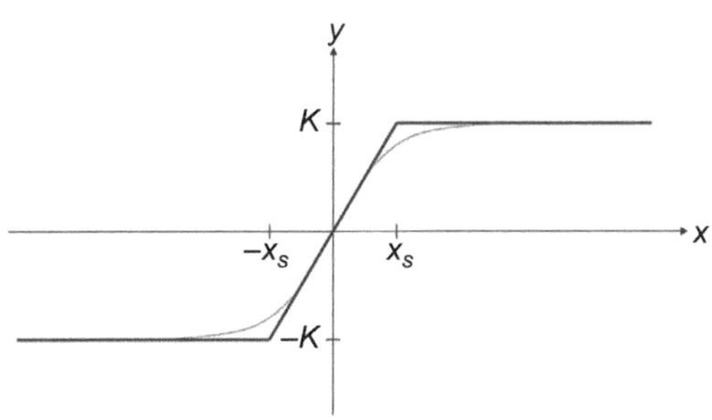

Figure 4.4 The input–output characteristic of the saturation. The black line shows the ideal characteristic, and the gray curve is closer to typical actual behavior.

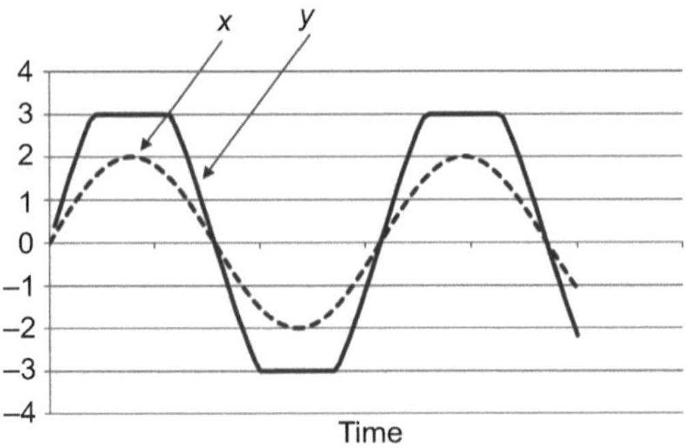

Figure 4.5 The output of the saturation element with a gain of 2 and a limit of ±3.

the modeled behavior. Often the saturation is modeled as a hard saturation, meaning that there is an abrupt transition from the linear to saturation region. However, in real systems, the saturation is usually more accurately modeled by a curve smoothly connecting the linear and saturation regions. The smooth transition can be modeled mathematically, but one must consider the benefit of adding complexity to the model and whether the expression in (4.4) is accurate enough. For example, the model can have a different level of simplicity if the desired use of the model is to accurately predict the output versus a model of the plant in a control system.

Figure 4.5 shows the output of the block as a function of time when the input is a sinusoid. In this case, the gain of the linear region is 2, and the saturation values are ±3.

Systems with saturation nonlinearities are very common in hardware applications.[1] They appear in systems that "run out of room." Opamps, transistors, springs, and motors are examples of such systems. The opamp example is discussed in Section 1.2.2. When dealing with a system exhibiting saturation, a common approach is to design the system so that it operates within the linear region so that linear techniques can be used.

4.1.3 Dead Zone

The characteristic of a **dead zone** is mathematically expressed as

$$y = \begin{cases} a(x+d) & x < -d \\ 0 & -d \le x \le d \\ a(x-d) & x > d \end{cases} \tag{4.5}$$

[1] An entire book has been written on this topic. See Liu and Michel (1994).

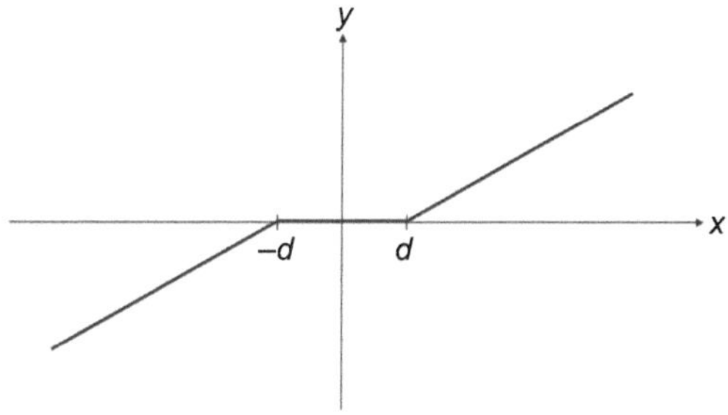

Figure 4.6 The input–output characteristic of the dead zone nonlinearity.

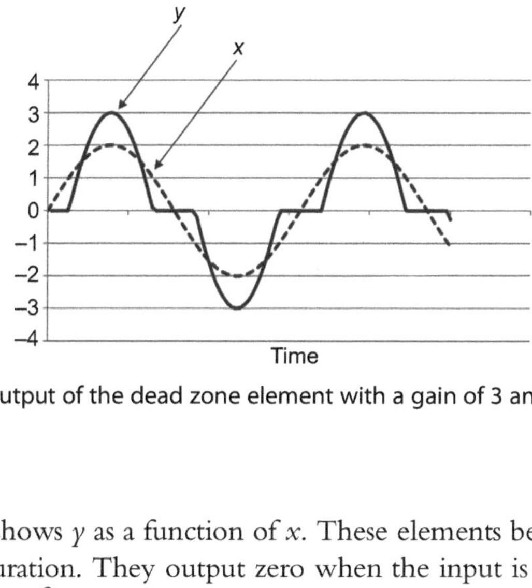

Figure 4.7 The output of the dead zone element with a gain of 3 and a threshold of 1.

Figure 4.6 shows y as a function of x. These elements behave almost the opposite of saturation. They output zero when the input is within a certain range and are linear[2] when the input exceeds some threshold. Figure 4.7 shows the output as a function of time when the input is a sinusoid. In this case, $a = 3$ and $d = 1$.

Dead zone nonlinearities appear in systems that get "stuck" and don't respond until the input is beyond some minimum value. Examples of this behavior can be found in direct current (DC) motors, hydraulic systems, steering systems, and limb motion in animals. In Section 4.2, we will investigate an actuator in a jet engine that exhibits a dead zone and how it affects the system performance.

[2] The term "linear" is being used loosely here. The function is not linear for $x > d$ because it does not go through the origin, but if one considers the input to be how far x is past d, that is $x - d$, then there is linear behavior.

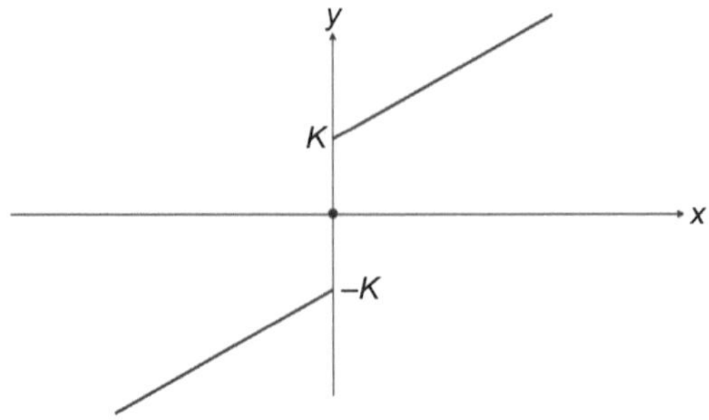

Figure 4.8 The input–output characteristic of the Coulomb and viscous friction nonlinearity.

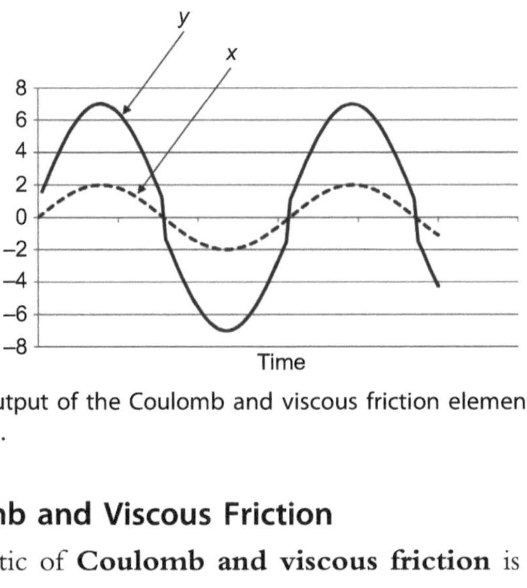

Figure 4.9 The output of the Coulomb and viscous friction element with a gain of 3 and an offset of 1.

4.1.4 Coulomb and Viscous Friction

The characteristic of **Coulomb and viscous friction** is mathematically expressed as

$$y = \begin{cases} ax - K & x < 0 \\ 0 & x = 0 \\ ax + K & x > 0 \end{cases} \tag{4.6}$$

Figure 4.8 shows y as a function of x. This element behaves linearly[3] for $x \neq 0$ and exhibits a discontinuity at $x = 0$. Figure 4.9 shows the output when the input is a sinusoid, with $a = 3$ and $K = 1$.

[3] Again the term "linear" is being used loosely. It is not truly linear because of the offset K. This type of relationship is known as an *affine transformation*.

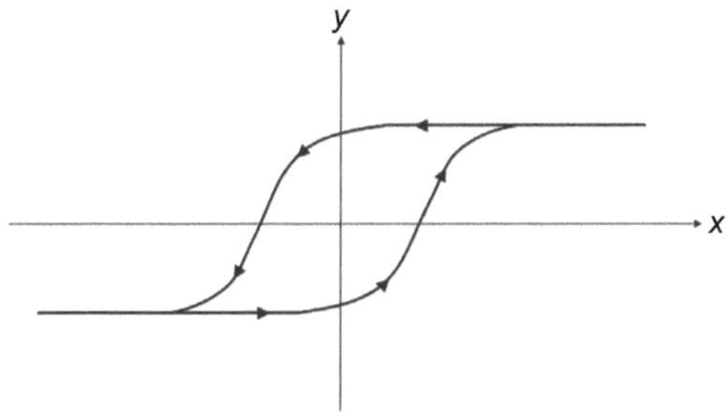

Figure 4.10 A typical hysteresis curve.

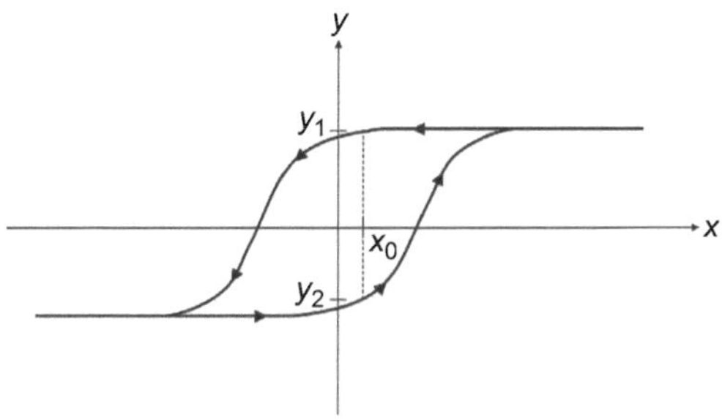

Figure 4.11 Demonstrating how to interpret a hysteresis curve.

The Coulomb and viscous friction nonlinearity is a common way to model friction. It accounts for static friction as manifested by the offset K and dynamic (or moving) friction through the slope a.

4.1.5 Hysteresis

Hysteresis is the most complicated of all the nonlinearities presented because y is not simply a function of x, as it has been in the other cases. Rather, y is also a function of $\dot{x}$. Unlike the other nonlinearities, there is no single mathematical expression for hysteresis.[4] It is typically expressed graphically. Figure 4.10 shows a typical example of a hysteresis curve.

The first thing to note about the relationship between x and y is that for a given x, there may be two possible values for y. The way to interpret the relationship is as follows. Consider a specific value for x shown in Figure 4.11 as x_0. The question is whether the y value will take on value y_1 or y_2. Just knowing x_0 is not enough information. The value of y depends on the where x came from. If x is increasing, then $y = y_2$. If x is decreasing,

[4] As with saturation, entire books have been written on the mathematics of hysteresis and where it is encountered in real systems. See, for example Mayergotz (2003), and Bertotti and Mayergotz (2006).

then $y = y_1$. The curve is followed in the direction given by the arrows and indicates whether to use the upper or lower part.

A system with hysteresis is one that has memory. Its output depends on where came from. Many physical phenomena display some form of hysteresis. Examples include backlash in gears caused by excess play, forces exerted by elastic materials, Schmitt triggers from electronic circuits, and magnetization of ferrous materials.

Hysteresis may be a desirable or undesirable characteristic of a system. Hysteresis may be intentionally designed into a system to reduce sensitivity to noise or time lag. An example of such a design is a thermostat that has different temperature thresholds for turning the heater on and off. If the desired temperature is set to 70 °F, then the switching does not actually occur at 70 °F because any noise in the temperature sensor could cause the system to repeatedly switch on and off very quickly. Instead, the thresholds may be set at 68 °F and 72 °F. The thresholds should be set far enough apart so that sensor noise does not cause the high-frequency on/off signals. If the room is heating, the heater turns off at 72 °F, and if the room is cooling, the heater turns on at 68 °F. As such, any error in temperature measurement less than 4 °F would not cause on/off oscillations. Schmitt triggers may be used as the circuit control in these control systems.

An example of undesirable hysteresis can be found as a source of error in some types of sensors. Hysteresis is common in temperature sensors because the material doing the sensing takes time to heat or cool. This time lag results in a different temperature value if the sensor is heating or cooling. The error introduced by this effect is usually given in the datasheet of the device as a worst-case offset.

4.2 LIMIT CYCLES

One interesting behavior that nonlinear systems can exhibit is a limit cycle. A **limit cycle** is an oscillation that has a fixed amplitude and frequency regardless of initial conditions, external inputs, or disturbances. Although linear systems may oscillate, limit cycles are specific to nonlinear systems.

Let us formalize the definition of an oscillation. A system trajectory oscillates if there exists a $T > 0$ such that $x(t + T) = x(t)$ for all $t \geq 0$. The period of oscillation is T. A phase plot of such a trajectory has a closed curve and is thus sometimes called a periodic orbit or a closed orbit.

First let's consider a linear system that oscillates. Figure 4.12 shows such a harmonic oscillator, a mass connected to a wall through a spring with no

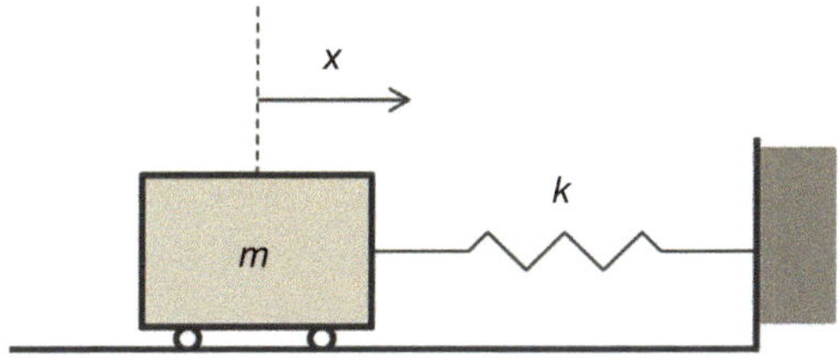

Figure 4.12 A mass-spring system that exhibits oscillations.

damping or friction. If the mass is held at a position other than equilibrium and let go, the mass will continue to move back and forth forever.

The state space model of this system is

$$\begin{bmatrix} \dot{x} \\ \dot{x}_2 \end{bmatrix} = \begin{bmatrix} 0 & 1 \\ -\dfrac{k}{m} & 0 \end{bmatrix} \begin{bmatrix} x_1 \\ x_2 \end{bmatrix} \tag{4.7}$$

where $x_1 = x$ and $x_2 = \dot{x}$. The eigenvalues are $\pm j\sqrt{\frac{k}{m}}$. Systems with imaginary eigenvalues exhibit oscillations. However, the amplitude of this oscillation depends on the initial x position where the mass was let go. Additionally, if there is any disturbance to the system, such as a small change in a parameter value, the oscillation will change because it is not structurally stable. Structural instability means that a small change in the parameter can move the eigenvalue off the imaginary to either the left or right half of the complex plane resulting in stable or unstable behavior respectively. When this happens either the oscillations dampen out or the trajectory goes off to infinity.

We also saw oscillations in the undamped pendulum as shown in Figure 2.20. The pendulum is a nonlinear system, so is this a limit cycle? No, because the period of oscillation depends on the initial angle of release. The pendulum will swing back and forth to the amplitude from which it was released. The oscillation depends very much on the initial conditions. Also, if an external torque is applied, the pendulum can be brought to a stop. If this trajectory were a limit cycle, the oscillations would continue even with the external influence.

For limit cycles to occur, there needs to be a fundamental difference in the system compared with the mass-spring or pendulum. It is this fundamental difference that allows these oscillations to appear no matter what. Even when external inputs are applied to the system, the oscillation will superimpose itself on the nominal output.

Rather than discuss the theoretical foundations of limit cycles, we will show their existence and behavior through an example of a real-world system that exhibits them.[5]

4.2.1 Simulink Example: Limit Cycles in a Jet Engine Control System

We saw an example of a limit cycle in the mechanical belt system in Section 3.4.5.3. In this example, we present another situation in which a limit cycle occurs—the control system for a jet engine that has a dead zone nonlinearity. The block diagram of the system is shown in Figure 4.13.

At the heart of the system is the actuator, which is a cantilevered device that allows fuel to pass into the engine. When electric current passes through the actuator, it causes deflection. However, the deflection doesn't start until a minimum threshold current is reached. When this level is exceeded, the lever's movement is proportional to the current. For this reason, the actuator is modeled by a dead zone nonlinearity. The lever's movement is integrated (the $\frac{1}{s}$ block) to obtain its position, which is equivalent to fuel flow in this model. The inner loop maintains the desired fuel flow through negative feedback. The fuel then flows into the engine, which is modeled by a first order low pass filter, and the resulting output is the engine rotational speed ω. The outer loop uses a proportional-integral (PI) controller, which is used to match the output ω to the desired engine speed ω_d.

The Simulink model used to simulate this system is shown in Figure 4.14. At the output, the engine speed is plotted versus time by the block titled "Output." The derivative of the output is taken, and together with the output, is plotted on an XY plot to generate the phase plot. For this simulation, the values for the PI controller were $K_p = 3$ and $K_i = 10$, and the dead zone had its deadband set to ± 1.5.

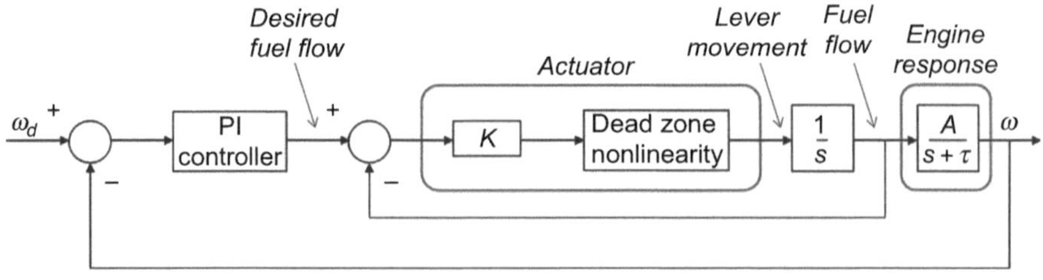

Figure 4.13 Block diagram for the engine control system. *PI*, proportional-integral.

[5] Interested readers can investigate limit cycle theory in Khalil (1996) and Vidyasagar (1993).

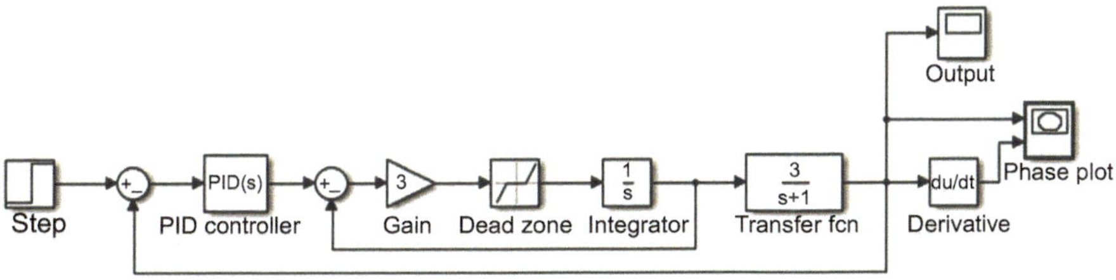

Figure 4.14 Simulink program for the engine control system. *PID,* proportional-integral-derivative.

Figure 4.15(a), shows the response of the system when the desired output signal is a step of height 1. There is an oscillation appearing that has a frequency of about 0.7 Hz and an amplitude of about 0.5 sitting on top of the desired output of 1. Figure 4.15(b), shows the corresponding phase plot with the derivative of the engine speed plotted against the engine speed. As with the mechanical belt system, there is an oscillating orbit to which the trajectory converges. This oscillation is the limit cycle.

Compare Figure 4.15 with Figure 4.16 in which the dead zone element is eliminated. In this case, the system behaves as expected and as desired. After a transient that stops at approximately 15 s, the output settles to a value of 1, matching the desired input. This behavior is shown another way in the phase plot in Figure 4.16(b). The trajectory converges to the point (1, 0) representing an output speed of 1 with no acceleration.

The limit cycle is always present in this system (although it may go away for different values of PI controller gains, which would need to

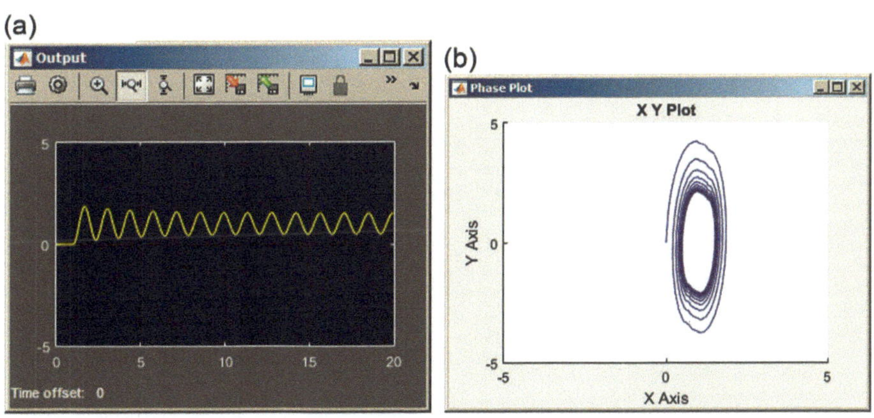

Figure 4.15 The engine response to a step input. (a) The engine speed as a function of time. (b) The phase plot for the system.

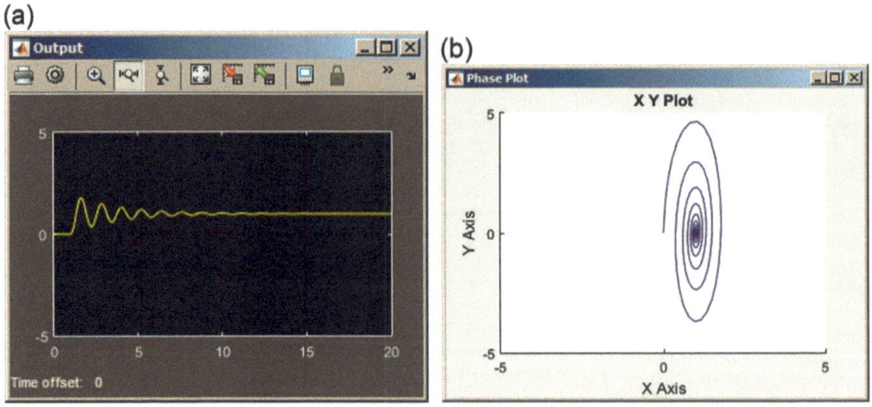

Figure 4.16 The ideal system without a dead zone element. (a) The engine speed as a function of time. (b) The phase plot for the system.

be tested). Whatever signal for ω_d is input to the system, the output ω will converge to it because of the feedback control system, but the limit cycle will be present on top of that signal. As shown in the figures, the step response in Figure 4.15(a), matches Figure 4.16(b), except the oscillation is added. This idea is further demonstrated in Figure 4.17. In Figure 4.17(a), the input signal is a sinusoid of amplitude 2 and frequency of approximately 0.05 Hz. The output matches the input, but the smaller amplitude and higher frequency oscillation is superimposed on it. Also in Figure 4.17(b), the input is a ramp, and the output is seen to track it with a limit cycle of the same amplitude and frequency added to it.

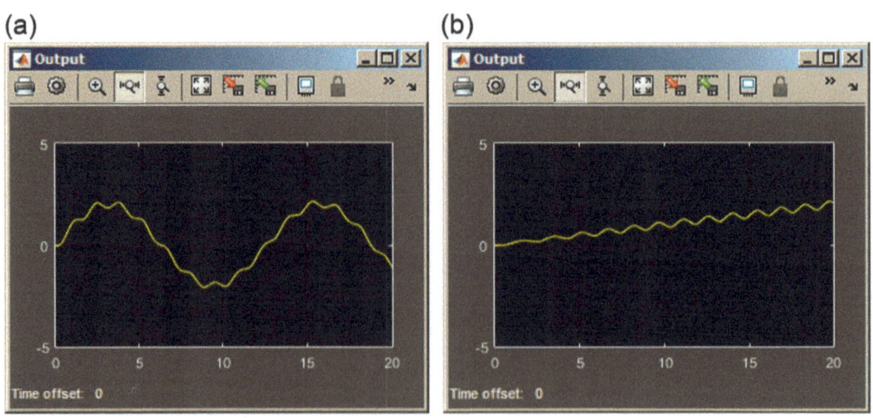

Figure 4.17 The engine response when the input is sinusoidal (a) and a ramp (b).

The limit cycle is indeed interesting system behavior, but in practice, it may be undesirable. Certainly, in this example, it is. This limit cycle represents an oscillation that is present in the engine rotation. In an actual engine, the rotational speed would vary at a frequency of about 0.7 Hz even as the pilot tries to hold it steady. In general, this type of behavior may lead to inefficiencies in operation; unnecessary wear and tear on mechanical components that could lead to a decreased lifetime; or in the worst case, conditions that could compromise safety.

4.3 BIFURCATION

Bifurcation in a nonlinear system can be defined simply as a change in behavior resulting from a small change in a parameter. By behavior, we typically mean a change in the number equilibrium points, a change in the type of equilibrium points (stable or unstable), or the emergence of a limit cycle. By small change in a parameter, we mean that there is a threshold above which the system exhibits one type of behavior and below it exhibits another.

We will discuss bifurcation in the context of two examples. The first example is the logistic differential equation, and the second example returns to the mechanical belt system.

4.3.1 Example: Bifurcation in the Logistic Differential Equation

The logistic differential equation is a model of population dynamics. The model describes the growth of a single species (unlike the predator—prey system in Chapter 3). The assumption is that the population increases because of reproduction and decreases because of starvation and harvesting. The growth is modeled as being proportional to two factors: (1) the current population and (2) how far the current population is from its peak.

The logistic differential equation is

$$\dot{x} = rx(1 - x) - h \tag{4.8}$$

where x is the population density in the range [0, 1], r is the parameter describing growth rate, and h is the harvest rate.

The equilibrium points x^* can be found by setting (4.8) to zero and solving for x, resulting in

$$rx(1 - x) - h = 0 \tag{4.9}$$

The solution to (4.9) is

$$x^* = \frac{1}{2} \pm \sqrt{\frac{1}{4} - \frac{h}{r}} \tag{4.10}$$

Because x represents a population size, only real values can be considered for equilibrium points. Thus, there are three possibilities to consider:

i. $h > \frac{r}{4}$, no equilibrium points exist
ii. $h = \frac{r}{4}$, one equilibrium point exists
iii. $h < \frac{r}{4}$, two equilibrium points exist

As the parameter h changes, the number of equilibrium points changes, and the system bifurcates. Note that the system behavior also changes with r, but in a real-life scenario, it is more likely that we would have more control over h. A **bifurcation diagram** for this system is a plot of the equilibrium points versus h as shown in Figure 4.18. As can be seen, there are two equilibrium points for low h values, and as h increases, the equilibrium points get closer together until they become the same value at h_0. Then as h increases further, there are no equilibrium points. Mathematically, for $h > h_0$, the value of x would keep decreasing forever. However, in reality, the population would fall to zero and remain there (and thus be at an equilibrium point) because it cannot take on negative values.

Let's also investigate the stability of these equilibrium points. Because the system is first order, the plot of $\dot{x}$ versus x shown in Figure 4.19 will help us greatly to determine stability. In the figure, first focus on the equilibrium point at $x = 0.25$. To the left of this point, the derivative is negative, so any initial values starting just to the left of 0.25 will decrease and thus move away from that point. To the right of $x = 0.25$, the derivative is positive, so any initial values starting just to the right 0.25 will increase and also move away from that point. Based on the fact that any

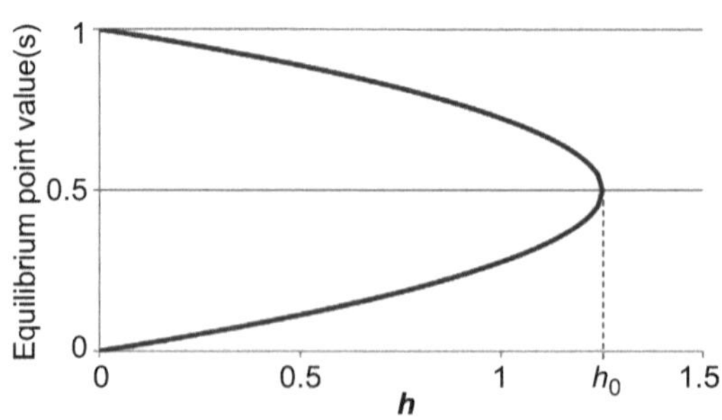

Figure 4.18 Bifurcation diagram for the logistic differential equation with $r = 5$.

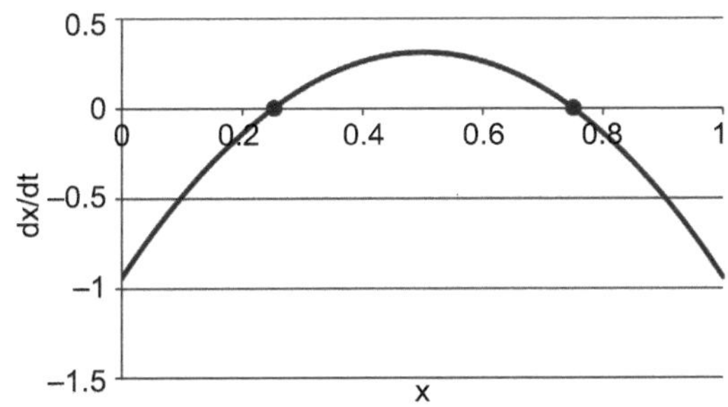

Figure 4.19 Plot used to determine the stability of equilibrium points.

points starting near $x = 0.25$ will move away from it, we can conclude that this equilibrium point is unstable.

Now consider the equilibrium point at $x = 0.75$. To the left of this point, the derivative is positive, so initial values starting just to the left of 0.75 will increase and move toward it. Similarly, the derivative to the right of 0.75 is negative, so initial values starting in that region will decrease back toward that point. Because any point starting near $x = 0.75$ will be attracted toward it, we can conclude that this equilibrium point is asymptotically stable.

We can verify these results by looking at a Lyapunov function and applying Lyapunov's theorem for stability. For $x^* = 0.75$, define the Lyapunov function as

$$V(x) = (x - 0.75)^2 \tag{4.11}$$

which is positive except for $V(0.75) = 0$. Then taking the derivative gives

$$\dot{V}(x) = 2(x - 0.75)\dot{x} \tag{4.12}$$

When $x < 0.75$, then $x - 0.75 < 0$ and $\dot{x} > 0$, thus $\dot{V}(x) < 0$. When $x > 0.75$, then $x - 0.75 > 0$ and $\dot{x} < 0$ and thus again $\dot{V}(x) < 0$. Note that we are only considering signs of $V(x)$ and $\dot{V}(x)$ for x near 0.75. Because $V(x)$ satisfies the conditions of Lyapunov's stability theorem, we can conclude (again) that the equilibrium point at $x^* = 0.75$ is asymptotically stable.

Often the bifurcation diagram includes arrows indicating how points near the equilibrium point move and gives an indication of stability. Figure 4.20 shows the bifurcation diagram for the logistic differential equation modified to include this information. For large h, there is no equilibrium point, and the population will tend toward zero as shown. This makes intuitive sense because if the harvest rate is high enough, the species will eventually die out.

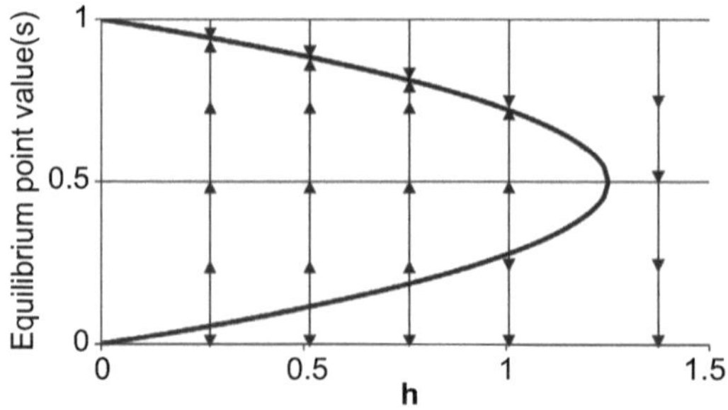

Figure 4.20 Bifurcation diagram showing direction of travel for x^*.

4.3.2 MATLAB Example: Bifurcation in the Mechanical Belt System

Let us return to the mechanical belt system that showed a limit cycle in Section 3.4.5.3. However, now the equations of motion will remain in the original coordinates: mass displacement x and velocity $\dot{x}$. The state equations for this system are

$$\begin{aligned}\dot{x}_1 &= x_2\\ \dot{x}_2 &= -\frac{k}{m}x_1 - \frac{b}{m}x_2 + \frac{1}{m}\text{sgn}(x_2 - v_0)\end{aligned} \tag{4.13}$$

This system has an equilibrium point at $x_1 = \frac{\text{sgn}(-v_0)}{k}$ and $x_2 = 0$. The MATLAB code to simulate the system is shown below.

```
% mechanical_belt_bifurcation.m

% Close all figures and clear all variables
close all
clear all

% How long to simulate (in seconds)
t_end = 20;

% Set initial conditions on the system
x1_0 = 1;
x2_0 = 0;

% Solve the system equations
[T X] = ode45(@mechanical_belt_model,[0 t_end],[x1_0 x2_0]);

% Save the results in a vector
x1 = X(:,1);
x2 = X(:,2);
```

```
% Plot the results
plot(T,x1,'b')
xlabel('Time (seconds)')
ylabel('x_1')
figure
plot(x1,x2,'b',x1_0,x2_0,'bo')
xlabel('x_1')
ylabel('x_2')
```

The file that defines the model is below.

```
function dx = mechanical_belt_model(t,x)

% Define the model parameters
m = 1;
k = 10;
b = 2;
v0 = -1;

dx = zeros(size(x));

% Define the dynamical equations
dx(1) = x(2);
dx(2) = -(k/m)*x(1)-(b/m)*x(2)+(1/m)*sign(x(2)-v0);
```

Figure 4.21 shows the response when $v_0 = -1$. The equilibrium point shows asymptotically stable behavior. The initial condition $x_1 = 1$ and $x_2 = 0$ settles to the equilibrium at $x_1 = 0.1$ and $x_2 = 0$. Figure 4.22 shows the response when $v_0 = -0.1$. This response is quite different compared with Figure 4.21 because there is now a limit cycle, and the mass position oscillates around the origin. This is an example of bifurcation because there is a change in behavior. Somewhere between $v_0 = -1$ and $v_0 = -0.1$, the equilibrium point went from being asymptotically stable to unstable, and a limit cycle emerged.

4.4 CHAOS

Another interesting behavior that can occur in nonlinear system is **chaos**. The common definition of a chaotic system is that it has sensitive dependence on initial conditions. To put it another way, if we start a

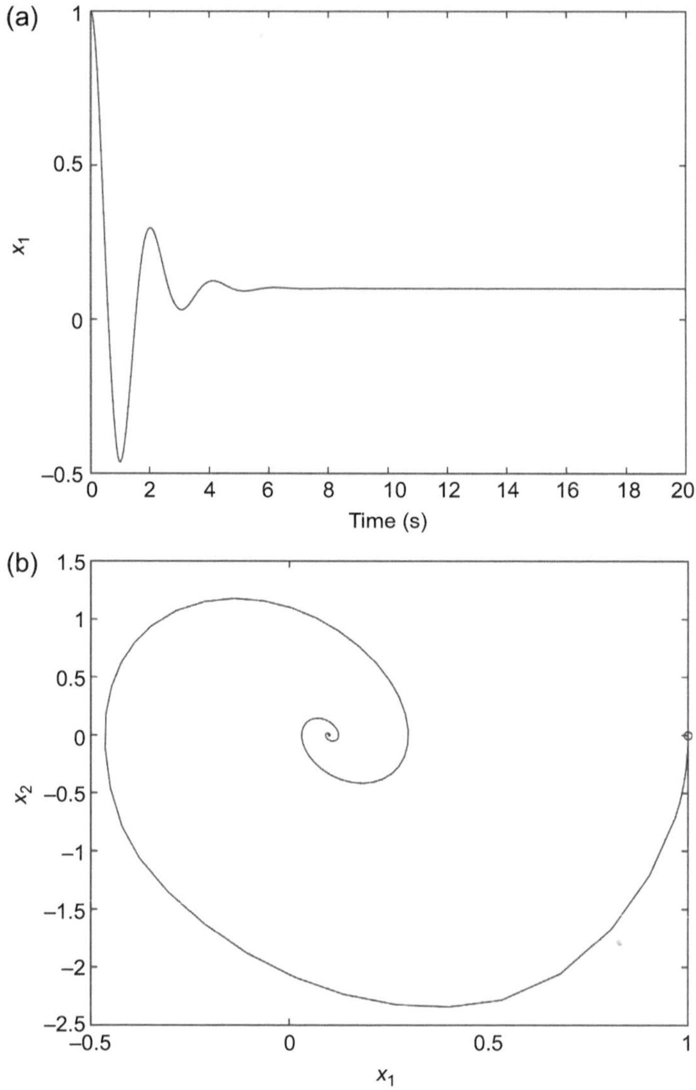

Figure 4.21 The response of the mechanical belt system when $v_0 = -1$. The mass position versus time (a) and the phase plot (b).

system at two different initial conditions, say x_{01} and x_{02}, then the trajectories resulting from each initial condition may be widely different from each other. Figure 4.23 shows the phase plot of such a system. Contrast the behavior with linear systems, which exhibit the scaling and additive properties. In linear systems, two initial conditions that start

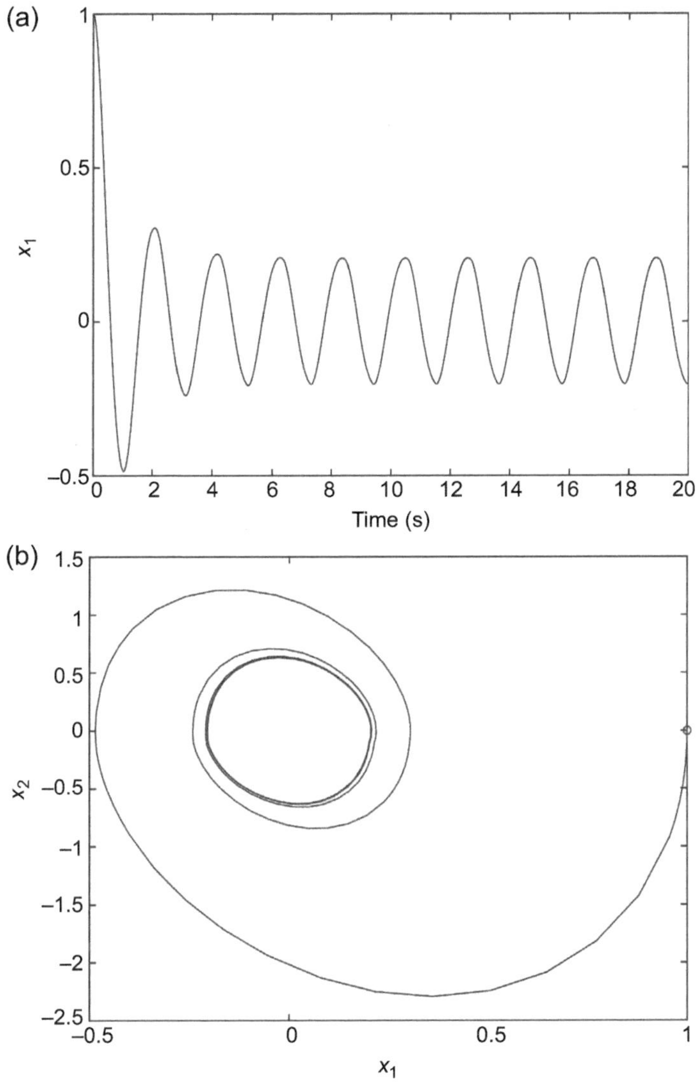

Figure 4.22 The response of the mechanical belt system when $v_0 = -0.1$. The mass position versus time (a) and the phase plot (b).

close to each other will have trajectories that behave similarly and stay relatively close to each other.

Chaos is often described by the butterfly effect—a butterfly flapping its wings causes a hurricane on the other side of the world. The relatively small amplitude of butterfly wings is equivalent to a small change in initial

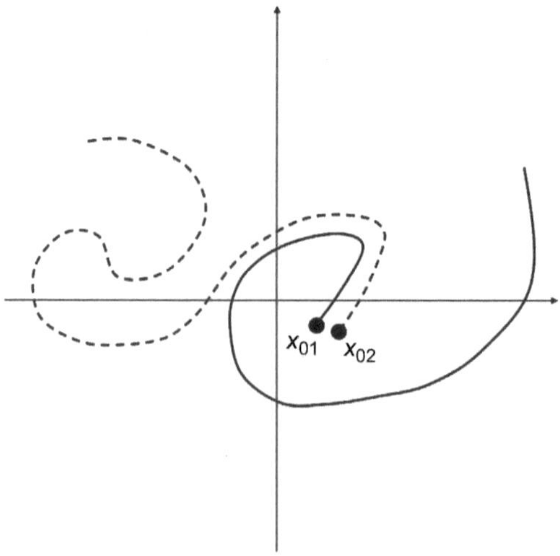

Figure 4.23 Trajectories in a chaotic system may diverge from each other even if the initial conditions are close.

condition. Surely a butterfly can't have much of an effect on atmospheric conditions! But even this small change is enough to make the difference between a nice sunny day and a storm (weather trajectory) in another part of the world.

One of the key ideas with chaotic systems is that they are unpredictable. Even though a system may be deterministic and the model may be 100% accurate, the long-term behavior simply cannot be predicted. Let's explore this idea in the context of the logistic equation.

4.4.1 Example: Chaotic Behavior in the Logistic Equation

The logistic equation is a discrete-time version of the logistic differential equation discussed in the previous section. The logistic equation takes the form

$$x[n+1] = rx[n]\big(1 - x[n]\big) \tag{4.14}$$

where r is the growth rate parameter, x represents population density and has range [0, 1], and n is a discrete time interval (days, years, generations, and so on).

Figure 4.24 shows x versus n for several different values of r. In each case, two trajectories are plotted, one starting at $x = 0.2$ and the other starting at $x = 0.3$. As r increases, some interesting behavior emerges. For $r = 2.7$, the trajectories both converge to a steady value after a transient. In this case, the system is stable as any initial value will converge to the same

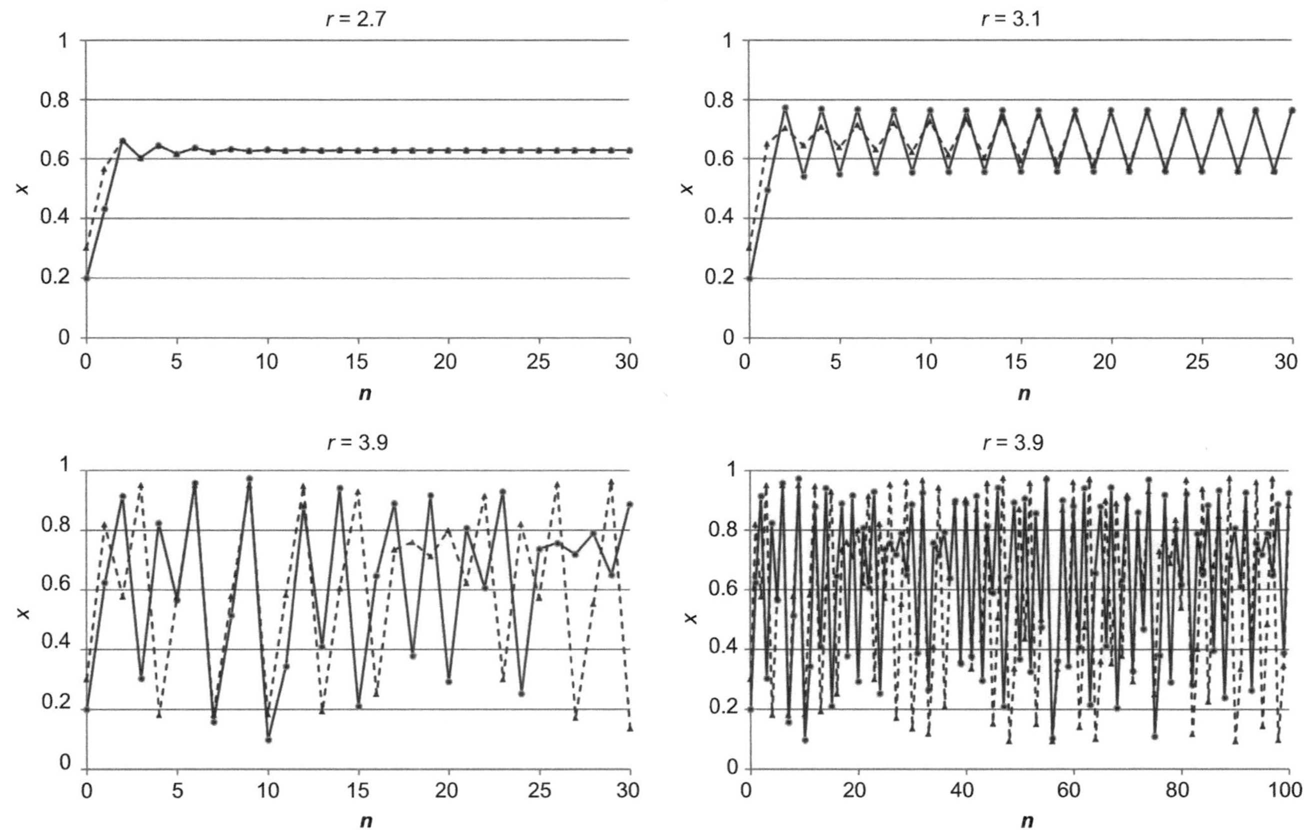

Figure 4.24 The emergence of chaotic behavior in the logistic equation as growth rates increase. Two trajectories are plotted with different initial conditions: $x = 0.2$ (solid line) and $x = 0.3$ (dashed line).

steady-state value. Similarly, for $r = 3.1$, the trajectories come together after a transient. However, the steady-state behavior is not a single value but a cyclic signal, known as a 2-cycle because it takes two steps in n to return to the same value. This is an example of a stable limit cycle because all initial conditions converge to this 2-cycle. Somewhere between $r = 2.7$ and $r = 3.1$, the system underwent a bifurcation.

For $r = 3.9$, the system exhibits a completely different behavior. In this case, the trajectory doesn't settle to any observable pattern, even when the values are plotted for n up to 100. But that isn't the only unusual behavior. Notice that the two trajectories are not similar at all. This is an example of sensitive dependence on initial conditions. One could argue that 0.2 and 0.3 are not very close. However, if the two initial conditions were started even closer to each other, say 0.2 and 0.201, the trajectories would still behave quite differently. In fact, making the initial condition differ by 0.0000000001 results in divergent trajectories; it just takes a larger number of time intervals to start seeing the difference.

The logistic equation is quite interesting. The same system, governed by the same modeling equation, displays three distinct behaviors: asymptotic stability, limit cycling, and chaos.

Chaos can have serious implications in engineering systems, and it is important to know this behavior exists. How many engineers have spent endless hours in lab running and rerunning experiments thinking something was wrong with the design or the experimental setup when in fact the bizarre behavior was inherent in the system itself?

As important as it is to understand that chaos might manifest itself as undesirable behavior, it is also possible to harness its power for system design. Next we explore a control algorithm that uses chaos to ensure favorable robotic behavior.

4.4.2 MATLAB Example: Using Chaos to Control a Mobile Robot

In this example[6] we investigate a mobile robot that uses the Arnold equation to create trajectories to cover a given space. This type of robot behavior is desirable in settings such as cleaning or patrolling a certain area.

The robot configuration is shown in Figure 4.25. It is a two-wheeled differential drive robot located at position (x, y) in the plane, which makes an angle θ with the x-axis.

[6] For full details, see Nakamura and Sekiguchi (2001).

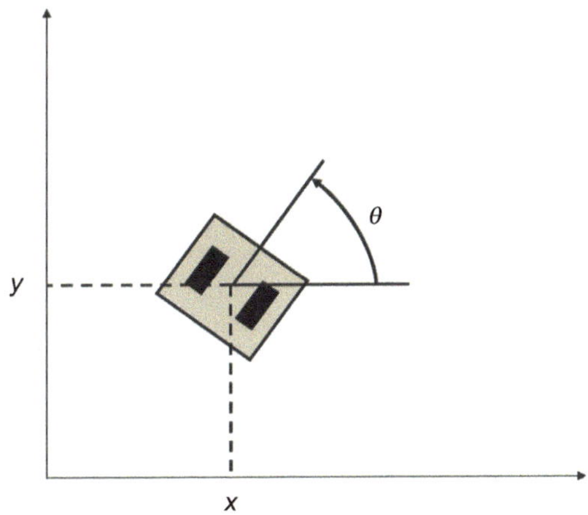

Figure 4.25 The robot configuration.

The robot is modeled kinematically, and it is assumed the linear velocity is only in the direction of the wheels (no perpendicular slippage). The equations of motion are

$$\begin{bmatrix} \dot{x} \\ \dot{y} \\ \dot{\theta} \end{bmatrix} = \begin{bmatrix} v \cos \theta \\ v \sin \theta \\ \omega \end{bmatrix} \tag{4.15}$$

where the linear velocity v and rotational velocity ω are inputs.

The Arnold equation is

$$\begin{bmatrix} \dot{x}_1 \\ \dot{x}_2 \\ \dot{x}_3 \end{bmatrix} = \begin{bmatrix} A \sin x_3 + C \cos x_2 \\ B \sin x_1 + A \cos x_3 \\ C \sin x_2 + B \cos x_1 \end{bmatrix} \tag{4.16}$$

This system of equations is known to exhibit chaotic behavior for certain values of A, B, and C. We can incorporate (4.15) and (4.16) together if we define x_3 to be θ. Then the system becomes fifth order with the following equations.

$$\begin{bmatrix} \dot{x}_1 \\ \dot{x}_2 \\ \dot{x}_3 \\ \dot{x} \\ \dot{y} \end{bmatrix} = \begin{bmatrix} A \sin x_3 + C \cos x_2 \\ B \sin x_1 + A \cos x_3 \\ C \sin x_2 + B \cos x_1 \\ v \cos x_3 \\ v \sin x_3 \end{bmatrix} \tag{4.17}$$

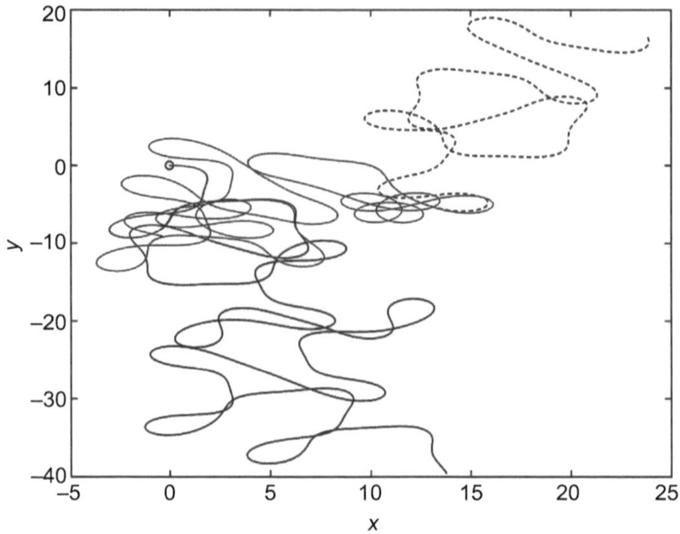

Figure 4.26 Three different trajectories starting near (0, 0) take different paths.

where A, B, and C are parameters to be chosen to achieve chaos and v is the linear velocity of the robot.

The system in (4.17) is chaotic because it exhibits sensitive dependence on initial conditions. Figure 4.26 shows three different trajectories starting near the origin. Although the initial conditions are near each other, the long-term behavior is quite different.

The robot's behavior in a room is simulated using the code below.

```
% robot_chaos.m

% Close all figures and clear all variables
close all
clear all

% Define simulation time
dt = 0.01;
t_end = 600;
t = [0:dt:t_end];

% Define room parameters
xmin = -1;
xmax = 1;
ymin = -1;
ymax = 1;
```

```
% Set initial conditions on the system
x0(1) = 0;
x0(2) = 0;
x0(3) = 0;
x0(4) = 0;
x0(5) = 0;

x = zeros(size(t));
y = zeros(size(t));

% Solve the system equations
for k = 1:length(t)
    [T X] = ode45(@robot_model,[0 dt],x0);

    x0 = X(end,:);

    % Adjust the angle if robot is at a wall
    if (x0(4) > xmax) || (x0(4) < xmin)
        x0(3) = pi-x0(3);
    end
    if (x0(5) > ymax) || (x0(5) < ymin)
        x0(3) = -x0(3);
    end

    % Save the results in a vector
    x(k) = x0(4);
    y(k) = x0(5);
end

% Plot the results
plot(x,y)
xlabel('x')
ylabel('y')
axis([xmin,xmax,ymin,ymax])
```

The robot model (4.16) is implemented in MATLAB code as the following.

```
function dx = robot_model(t,x)

% Define the model parameters
A = 1;
B = 0.5;
C = 0.5;
v = 1;

dx = zeros(size(x));
```

```
% Define the dynamical equations
dx(1) = A*sin(x(3))+C*cos(x(2));
dx(2) = B*sin(x(1))+A*cos(x(3));
dx(3) = C*sin(x(2))+B*cos(x(1));
dx(4) = v*cos(x(3));
dx(5) = v*sin(x(3));
```

The code above works similar to the example in Section 2.2.3.1, the computer-controlled vehicle dynamics, but for a different reason. Although the code in Section 2.2.3.1 simulated the sampling of the computer, in this case, we need to sample the robot's position to see if it is located at the wall. The following two `if` statements check where the robot is located. If it is at one of the walls, it rotates so that the angle of incidence equals the angle of reflection.

```
% Adjust the angle if robot is at a wall
if (x0(4) >= xmax) || (x0(4) <= xmin)
    x0(3) = pi-x0(3);
end
if (x0(5) >= ymax) || (x0(5) <= ymin)
    x0(3) = -x0(3);
end
```

Figure 4.27 shows the trajectory in a 2 m × 2 m room after it runs for 10 minutes. In Nakamura and Sekiguchi (2001), the authors claim that the chaotic algorithm performs better than a random walk algorithm in terms of coverage in a fixed amount of time because the robot can move at a constant velocity until it hits the boundary where the random robot had to stop and turn every 2 s.

4.5 LINEARIZATION

When encountering a nonlinear system, a typical first step is to try to linearize it. This is understandable because linear systems are so nice to work with, and many, many powerful tools are available from linear system theory. It is a lucky engineer whose nonlinear system can be well approximated by a linear one. In this section, we review two methods of linearization and explore the relationship between a nonlinear system and its linear approximation.

4.5.1 Linearization Using Taylor Series Expansion

A common method of linearizing a system is to use a truncated **Taylor series expansion**. Recall from calculus that any analytic function $f(x)$ can

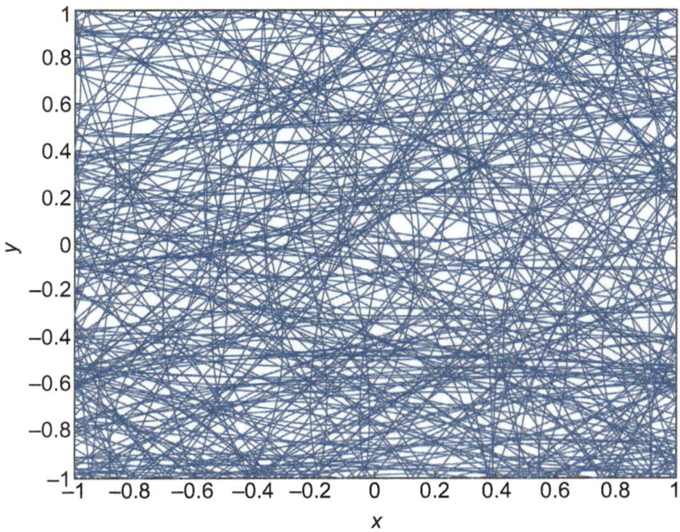

Figure 4.27 The robot's trajectory in a 2 m × 2 m room.

be represented by its Taylor series expansion, an infinite series involving every derivative of $f(x)$ evaluated at a point x_0. Then the series expresses the expansion of $f(x)$ about x_0.

Extending this idea to dynamical systems, assume the nonlinear time-invariant system is given by

$$\dot{x}(t) = f(x) \tag{4.18}$$

where x is an Nx1 vector and $f: \Re^N \rightarrow \Re^N$ is differentiable. The Taylor series expansion of f about a point x^* is

$$f(x) = f(x^*) + J(x^*)(x - x^*) + \cdots \tag{4.19}$$

where $J(x^*)$ is the Jacobian matrix defined as

$$J(x^*) = \begin{bmatrix} \dfrac{\partial f_1}{\partial x_1} & \cdots & \dfrac{\partial f_1}{\partial x_N} \\ \vdots & \ddots & \vdots \\ \dfrac{\partial f_N}{\partial x_1} & \cdots & \dfrac{\partial f_N}{\partial x_N} \end{bmatrix}_{x=x^*} \tag{4.20}$$

For linearization, the first two terms of (4.19) are kept, and the higher order terms are discarded.

4.5.1.1 Example: Linearizing the Pendulum

Let us return to the pendulum system given by

$$\begin{bmatrix} \dot{x}_1 \\ \dot{x}_2 \end{bmatrix} = \begin{bmatrix} x_2 \\ -\frac{g}{l}\sin x_1 - \frac{b}{ml^2}x_2 \end{bmatrix} = \begin{bmatrix} f_1(x_1, x_2) \\ f_2(x_1, x_2) \end{bmatrix} \tag{4.21}$$

where x_1 is the angle, x_2 is the angular velocity, g is acceleration caused by gravity, l is the pendulum length, m is the mass, and b is the friction coefficient. Using the truncated Taylor series expansion (4.19), we can linearize the system about $[0\ 0]^T$, and it then becomes

$$\begin{aligned} \begin{bmatrix} \dot{x}_1 \\ \dot{x}_2 \end{bmatrix} &\approx \begin{bmatrix} f_1(0,0) \\ f_2(0,0) \end{bmatrix} + \begin{bmatrix} 0 & 1 \\ -\frac{g}{l}\cos(0) & -\frac{b}{ml^2} \end{bmatrix} \begin{bmatrix} x_1 \\ x_2 \end{bmatrix} \\ &= \begin{bmatrix} 0 & 1 \\ -\frac{g}{l} & -\frac{b}{ml^2} \end{bmatrix} \begin{bmatrix} x_1 \\ x_2 \end{bmatrix} \end{aligned} \tag{4.22}$$

Similarly, we can linearize about the point $[\pi\ 0]^T$. Applying (4.19) again gives the linear system

$$\begin{aligned} \begin{bmatrix} \dot{x}_1 \\ \dot{x}_2 \end{bmatrix} &\approx \begin{bmatrix} f_1(\pi,0) \\ f_2(\pi,0) \end{bmatrix} + \begin{bmatrix} 0 & 1 \\ -\frac{g}{l}\cos(\pi) & -\frac{b}{ml^2} \end{bmatrix} \begin{bmatrix} x_1 \\ x_2 \end{bmatrix} \\ &= \begin{bmatrix} 0 & 1 \\ \frac{g}{l} & -\frac{b}{ml^2} \end{bmatrix} \begin{bmatrix} x_1 \\ x_2 \end{bmatrix} \end{aligned} \tag{4.23}$$

The final result in (4.23) is of the form $\dot{x} = Ax$.

4.5.1.2 Example: Linearizing a Friction Function

As discussed in Section 3.4.5.3, the friction coefficient between a mass and a belt can be modeled by the difference in their velocities $\dot{x} - v_0$, where $\dot{x}$ is the velocity of the mass with respect to an inertial frame and v_0 is the velocity of the belt. In other words,

$$g(\dot{x}, v_0) = f(\dot{x} - v_0) \tag{4.24}$$

Expanding g about v_0 using (4.19) gives

$$
\begin{aligned}
g(\dot{x}, v_0) &\approx g(0, v_0) + \left[\frac{\partial g}{\partial \dot{x}} \quad \frac{\partial g}{\partial v_0}\right]\Bigg|_{\substack{\dot{x}=0\\ v_0=v_0}} \begin{bmatrix} \dot{x}-0 \\ v_0 - v_0 \end{bmatrix} \\
&= g(0, v_0) + \frac{\partial g}{\partial \dot{x}}\bigg|_{\dot{x}=0} \dot{x}
\end{aligned} \tag{4.25}
$$

Substituting in f gives

$$
\begin{aligned}
f(\dot{x} - v_0) &\approx f(-v_0) + \frac{\partial f}{\partial \dot{x}}\bigg|_{\dot{x}=0} \dot{x} \\
&= f(-v_0) - \frac{\partial f}{\partial v_0}\bigg|_{\dot{x}=0} \dot{x}
\end{aligned} \tag{4.26}
$$

The substitution $\frac{\partial f}{\partial \dot{x}} = -\frac{\partial f}{\partial v_0}$ comes from the chain rule. If we assume that $\Delta v = \dot{x} - v_0$, then

$$
\begin{aligned}
\frac{\partial f(\Delta v)}{\partial \dot{x}} &= \left(\frac{\partial f(\Delta v)}{\partial v_0}\right)\left(\frac{\partial v_0}{\partial \Delta v}\right)\left(\frac{\partial \Delta v}{\partial \dot{x}}\right) \\
&= \left(\frac{\partial f(\Delta v)}{\partial v_0}\right)(-1)(1)
\end{aligned}
$$

Note a key difference between the previous two examples. In the case of the pendulum, the Taylor series expansion was about an equilibrium point (either $[0\ 0]^T$ or $[\pi\ 0]^T$). Then by definition $f(x^*) = 0$, so the first term of the series is eliminated, and what's left is a matrix multiplying x, resulting in the linear system $\dot{x} = Ax$. This will always happen when linearizing about an equilibrium point.

With the friction function, the situation is different. We linearized a *function* around an operating point v_0. The Taylor series expansion doesn't lose its first term, and the result is of the form $f(x) = ax + b$. This is *not* a linear system, as has been discussed before, because it does not pass through the origin. What this procedure did was take a nonlinear function and approximate it by a straight line at a certain point. However, the procedure for linearization was the same in both cases.

4.5.2 Linearization and Stability

In this section, we explore the relationship between linearized systems and their stability. In particular, if the linearized version of the system is stable, what can we conclude about the original nonlinear system's stability?

Fortunately, there is a relatively simple relationship that is summarized in the following theorem.

Lyapunov's Linearization Theorem: *Let x^* be an equilibrium point for the Nth order system $\dot{x} = f(x)$ and let the linearized system given by $\dot{x} = Ax$ where*

$$A = \begin{bmatrix} \frac{\partial f_1}{\partial x_1} & \cdots & \frac{\partial f_1}{\partial x_N} \\ \vdots & \ddots & \vdots \\ \frac{\partial f_N}{\partial x_1} & \cdots & \frac{\partial f_N}{\partial x_N} \end{bmatrix}_{x=x^*} \tag{4.27}$$

If all *eigenvalues of A are in the left half of the complex plane, then x^* is an asymptotically stable equilibrium point in the nonlinear system.*

If any *eigenvalues of A are in the right half of the complex plane, then x^* is an unstable equilibrium point of the nonlinear system.*

To obtain the equivalent theorem for discrete-time systems, one simply replaces "left (right) half of the complex plane" with "inside (outside) the unit circle."

Note that no conclusion can be made if the linearized system has imaginary eigenvalues (corresponding to the marginally stable case). Also, the conclusion about stability of the equilibrium point is local, not global. This makes sense intuitively because linearization is inherently local. When the system moves away from the point about which the system was linearized, the approximation may become very different from the original system. As the system moves away from that point, the higher order terms of the series expansion (that were dropped from the expression) may become large and have an effect on the system dynamics.

4.5.3 Feedback Linearization

Another linearization technique used in feedback control systems is feedback linearization. Let's assume the system has a block diagram of the form shown in Figure 4.28. The plant is the part of the system that needs to be controlled, and the controller takes the desired input $r(t)$ and the system's

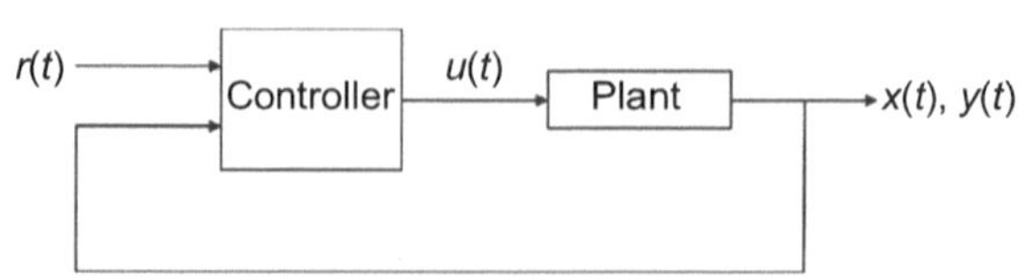

Figure 4.28 The structure of the feedback control system for feedback linearization.

state $x(t)$ and generates an appropriate input to the plant $u(t)$ to (hopefully) result in the overall system tracking the input $r(t)$.

Assume the plant is modeled by

$$\begin{aligned} \dot{x} &= f(x, u) \\ y &= g(x) \end{aligned} \tag{4.28}$$

where $x = [x_1, \ldots, x_N]$, $u = [u_1, \ldots, u_p]$, and $y = [y_1, \ldots, y_q]$.

The idea behind feedback linearization is to design the controller to cancel out nonlinearities in the plant so that the overall system in Figure 4.28 is linear. There are two general forms: **input-state linearization** and **input-output linearization**. For input-state linearization, it is assumed that all states $x_1, \ldots, x_N$ are available (either by direct measurement or calculation), and the goal is to cancel linearities in f so that the state equation is in linear form. Input—output linearization is similar except the goal is to design the controller to cancel linearities in g so that the output equation is in linear form and explicitly tied to the input u. We illustrate these concepts with two examples.

4.5.3.1 Example: Input-State Linearization of the Pendulum

Let us reconsider the pendulum system with applied torque T.

$$ml^2\ddot{\theta} = -mgl \sin \theta - b\dot{\theta} + T \tag{4.29}$$

Suppose we wish to design a controller so that the pendulum holds a certain constant angle θ_d. To stabilize the pendulum at $\theta_d \neq 0$, there needs to be an offset torque applied as shown in Figure 4.29.

Defining input u as

$$u = T - T_{off} \tag{4.30}$$

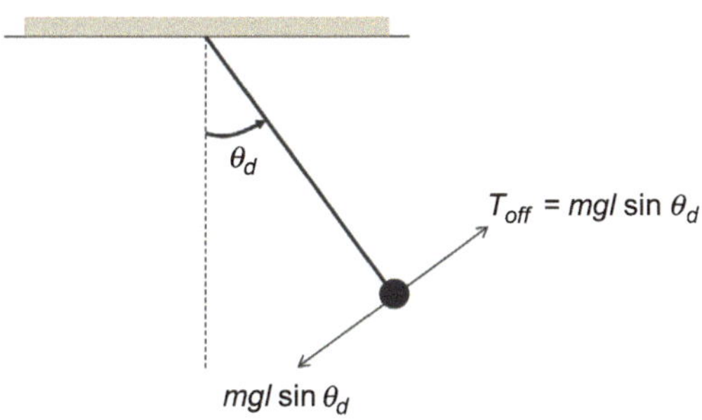

Figure 4.29 Holding the pendulum at a nonzero angle requires an offset torque.

the system equation (4.29) becomes

$$ml^2\ddot{\theta} = -mgl\sin\theta - b\dot{\theta} + u + mgl\sin\theta_d \tag{4.31}$$

First let's convert (4.31) to state-space form, but we won't use the usual states as $x_1 = \theta$ and $x_2 = \dot{\theta}$. Instead define the states as

$$\begin{aligned} x_1 &= \theta - \theta_d \\ x_2 &= \dot{\theta} \end{aligned} \tag{4.32}$$

Then the system equations become

$$\begin{aligned} \dot{x}_1 &= x_2 \\ \dot{x}_2 &= -\frac{g}{l}\sin(x_1 + \theta_d) - \frac{b}{ml^2}x_2 + \frac{1}{ml^2}u + \frac{g}{l}\sin\theta_d \end{aligned} \tag{4.33}$$

Note that there is an equilibrium point at $x_1 = x_2 = 0$, which is actually $\theta = \theta_d$ and $\dot{\theta} = 0$. If we choose u to cancel the nonlinear terms,

$$u = mgl(\sin(x_1 + \theta_d) - \sin\theta_d) + ml^2 v \tag{4.34}$$

then substituting (4.34) into (4.33) results in the linear system

$$\begin{aligned} \dot{x}_1 &= x_2 \\ \dot{x}_2 &= -\frac{b}{ml^2}x_2 + v \end{aligned} \tag{4.35}$$

Now one is free to choose v to satisfy whatever the system requirements are. In this case, we wish to have the pendulum go to $x_1 = x_2 = 0$. Then the control law

$$v = -k_1x_1 - k_2x_2 \tag{4.36}$$

will result in the system equations

$$\begin{aligned} \dot{x}_1 &= x_2 \\ \dot{x}_2 &= -k_1x_1 - \left(\frac{b}{ml^2} + k_2\right)x_2 \end{aligned} \tag{4.37}$$

Or in matrix form

$$\begin{bmatrix} \dot{x}_1 \\ \dot{x}_2 \end{bmatrix} = \begin{bmatrix} 0 & 1 \\ -k_1 & -\left(\frac{b}{ml^2} + k_2\right) \end{bmatrix} \begin{bmatrix} x_1 \\ x_2 \end{bmatrix} \tag{4.38}$$

Because the designer can choose k_1 and k_2, the eigenvalues of the system can be placed to give the desired system response. The actual input to the physical system T must be obtained through the transformation

$$\begin{aligned} T &= u + T_{off} \\ &= mgl(\sin(x_1 + \theta_d) - \sin\theta_d) + ml^2 v + mgl\sin\theta_d \\ &= mgl\sin\theta + ml^2 v \end{aligned} \tag{4.39}$$

4.5.3.2 Example: Input—Output Linearization of a Field-Controlled Direct Current Motor

Consider the schematic of the field-controlled DC motor shown in Figure 4.30.

The system equations for this motor are given by[7]

$$\begin{aligned} V_f &= R_f i_f + L_f \frac{di_f}{dt} \\ V_a &= R_a i_a + L_a \frac{di_a}{dt} + K_1 i_f \dot{\theta} \\ J\ddot{\theta} &= K_2 i_f i_a - K_3 \dot{\theta} \end{aligned} \tag{4.40}$$

where K_1, K_2, K_3, and J are constants for the motor. If we assume that $u = V_f$ is the input and V_a is held constant, defining the states as $x_1 = i_f$, $x_2 = i_a$, and $x_3 = \dot{\theta}$, the state equations of the motor are

$$\begin{aligned} \dot{x}_1 &= -\frac{R_f}{L_f} x_1 + \frac{1}{L_f} u \\ \dot{x}_2 &= -\frac{K_1}{L_a} x_1 x_3 - \frac{R_a}{L_a} x_2 + \frac{1}{L_a} V_a \\ \dot{x}_3 &= \frac{K_2}{J} x_1 x_2 - \frac{K_3}{J} x_3 \end{aligned} \tag{4.41}$$

If we take the output of the system to be the angular speed, then the output equation is

$$y = x_3 \tag{4.42}$$

The feature of this system that is undesirable is not that the output equation is nonlinear but that the input u does not show up explicitly.

[7] This example was inspired by an example in Khalil (1996).

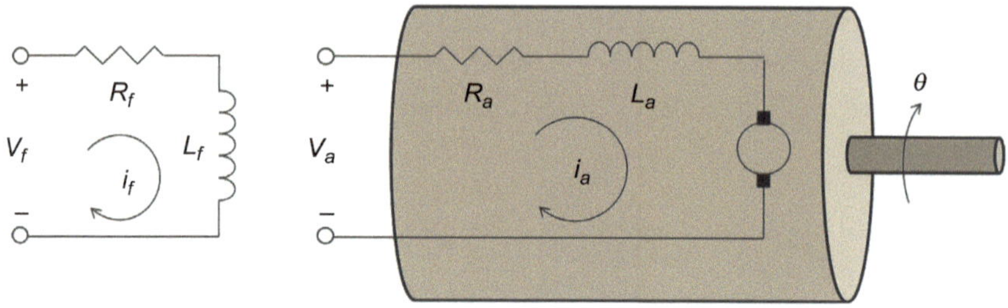

Figure 4.30 The schematic for a field-controlled direct current (DC) motor.

In this case, if we want to control the angular speed, it isn't obvious how to make that happen. Input—output linearization is a tool to make the control design task easier.

The procedure for input—output linearization is to take derivatives of the output y until the input u appears in the expression.

$$\begin{aligned}\dot{y} &= \dot{x}_3 \\ &= \frac{K_2}{J}x_1x_2 - \frac{K_3}{J}x_3\end{aligned} \tag{4.43}$$

$$\begin{aligned}\ddot{y} &= \frac{K_2}{J}(x_1\dot{x}_2 + \dot{x}_1x_2) - \frac{K_3}{J}\dot{x}_3 \\ &= -\frac{K_1K_2}{JL_a}x_1^2x_3 - \frac{K_2}{J}\left(\frac{R_a}{L_a}+\frac{R_f}{L_f}+\frac{K_3}{J}\right)x_1x_2 + \frac{K_2V_a}{JL_a}x_1 + \frac{K_3^2}{J^2}x_3 + \frac{K_2}{JL_f}x_2u\end{aligned} \tag{4.44}$$

Because it takes two derivatives for the input to appear, the system has **relative degree** 2.

The next step is to choose input u so that $\ddot{y} = v$ or specifically,

$$\begin{aligned}u &= \frac{JL_f}{K_2x_2}\left(\frac{K_1K_2}{JL_a}x_1^2x_3 + \frac{K_2}{J}\left(\frac{R_a}{L_a}+\frac{R_f}{L_f}+\frac{K_3}{J}\right)x_1x_2 - \frac{K_2V_a}{JL_a}x_1 - \frac{K_3^2}{J^2}x_3 + v\right) \\ &= \frac{K_1L_f}{L_a}\frac{x_1^2x_3}{x_2} + \left(\frac{R_aL_f}{L_a} + R_f + \frac{K_3L_f}{J}\right)x_1 - \frac{V_aL_f}{L_a}\frac{x_1}{x_2} + \frac{K_3^2L_f}{JK_2}\frac{x_3}{x_2} + \frac{JL_f}{K_2}\frac{1}{x_2}v\end{aligned} \tag{4.45}$$

With this substitution, the designer has direct control of y using v. However, one problem with this approach is that the **internal dynamics** of the system may be unstable. The internal dynamics of the system are the parts of the system that aren't seen and aren't part of the control design. In this case, the internal dynamics are governed by x_1 and x_2 whose state equations are now

$$\begin{aligned}\dot{x}_1 &= -\frac{R_f}{L_f}x_1 + \frac{1}{L_f}\left(\frac{K_1 L_f}{L_a}\frac{x_1^2 x_3}{x_2} + \left(\frac{R_a L_f}{L_a} + R_f + \frac{K_3 L_f}{J}\right)x_1\right. \\ &\qquad \left. - \frac{V_a L_f}{L_a}\frac{x_1}{x_2} + \frac{K_3^2 L_f}{J K_2}\frac{x_3}{x_2} + \frac{J L_f}{K_2}\frac{1}{x_2}v\right) \\ \dot{x}_2 &= \left(\frac{R_a}{L_a} + \frac{K_3}{J}\right)x_1 + \frac{K_1}{L_a}\frac{x_1^2 x_3}{x_2} - \frac{V_a}{L_a}\frac{x_1}{x_2} + \frac{K_3^2}{J K_2}\frac{x_3}{x_2} + \frac{J}{K_2}\frac{1}{x_2}v\end{aligned} \tag{4.46}$$

If the internal dynamics are unstable, x_1 and x_2 may grow without bound, meaning that the field current and armature current will increase and cause the motor to burn out. This may happen even if there is desirable output behavior. There is an entire theoretical basis for studying internal dynamics. Readers are encouraged to explore Khalil (1996) and Slotine and Li (1991) for further information.

Feedback linearization is unlike that obtained by truncated Taylor series expansion. The major difference is that feedback linearization is exact, not an approximation. It is merely transforming the model into a linear one by choosing an appropriate input, similar to transforming a physical model into its canonical counterpart by choosing appropriate state variables. However, this idea is also one of the drawbacks of feedback linearization. The technique requires precision in the model in order to cancel the nonlinearities. For example, in (4.34), exact measurement of the pendulum's mass and length as well as the gravitational constant are required.

This concludes our discussion of nonlinear system characteristics. Although not an exhaustive or mathematically rigorous study of these systems, some of the most interesting topics, such as limit cycles, bifurcation, and chaos, were presented. We examined the application of these concepts to jet engine control, population dynamics, mechanical belts, robotic control, DC motors, and pendulum systems. In moving from Chapter 3 to this chapter, we narrowed our focus from general dynamical systems to nonlinear systems. In moving to Chapter 5, we focus again on a specific type of system: Hamiltonian systems. Unlike our focus on nonlinear systems, this focus brings with it a very specific structure, as we will see in the next chapter.

REFERENCES

Bertotti, G., & Mayergotz, I. D. (2006). *The Science of Hysteresis* (Vol. 1). Oxford, UK: Elsevier.

Khalil, H. (1996). *Nonlinear Systems* (2nd ed.). Upper Saddle River, NJ: Prentice-Hall.

Liu, D., & Michel, A. N. (1994). *Dynamical Systems with Saturation Nonlinearities: Analysis and Design*. London: Springer-Verlag.

Mayergotz, I. (2003). *Mathematical Models of Hysteresis and Their Applications*. New York: Elsevier Science.

Nakamura, Y., & Sekiguchi, A. (2001, December). The chaotic mobile robot. *IEEE Transactions on Robotics and Automation, 17*(6), 898–904.

Slotine, J.-J., & Li, W. (1991). *Applied Nonlinear Control*. Englewood Cliffs, NJ: Prentice-Hall.

Vidyasagar, M. (1993). *Nonlinear Systems Analysis* (2nd ed.). Engelwood Cliffs, NJ: Prentice-Hall.

FURTHER READING

Franklin, G., Powell, J. D., & Emami-Naeini, A. (2010). *Feedback Control of Dynamic Systems* (6th ed.). Upper Saddle River, NJ: Pearson Higher Education.

Hassard, B., Kazarinoff, N. D., & Wan, Y.-H. (1981). *Theory and Applications of Hopf Bifurcation*. Cambridge, UK: Cambridge University Press.

Shaw, S., & Rand, R. H. (1989). The transition to chaos in a simple mechanical system. *International Journal of Nonlinear Mechanics, 24*(1), 41–56.

Simpson, D. (2010). *Bifurcations in Piecewise-Smooth Continuous Systems*. Hackensack, NJ: World Scientific Publishing Co.

CHAPTER 5

Hamiltonian Systems

5.1 OVERVIEW

In the earlier chapters, we encountered Hamiltonian systems, the pendulum being one example, but we did not view them from the perspective of the Hamiltonian function. We begin the discussion by first focusing on conservative dynamical systems and volume-preserving flows before formally defining Hamiltonian systems and presenting examples.

From a physics perspective, a conservative system is one in which the force F can be derived from the potential energy at a given position $V(x)$ by the relationship

$$F = -\frac{dV}{dx} \tag{5.1}$$

In such a case, the total energy (kinetic and potential) is constant along all possible trajectories. Contrast this with a dissipative system, in which energy is lost. Mathematically, a conservative system can be defined explicitly. Consider a time-invariant system modeled by

$$\dot{x} = f(x) \tag{5.2}$$

This system is a **conservative system** if there is some function $I(x)$, typically total energy, but it could be more a general quantity, such that

$$\frac{d}{dt} I(x) = 0 \tag{5.3}$$

The function $I(x)$ satisfying (5.3) is called an invariant. The meaning of (5.3) is that if you follow any trajectory of the system (5.2), the value of $I(x)$ will be the same at every point along that trajectory.

As a simple example of a conservative system, consider the model of an undamped harmonic oscillator.

$$\begin{aligned} \dot{x}_1 &= x_2 \\ \dot{x}_2 &= -ax_1 \end{aligned} \tag{5.4}$$

A Practical Approach to Dynamical Systems for Engineers
ISBN 978-0-08-100202-5
http://dx.doi.org/10.1016/B978-0-08-100202-5.00005-X

where $a > 0$. Define the function $I(x_1, x_2)$ by

$$I(x_1, x_2) = \frac{1}{2}\left(ax_1^2 + x_2^2\right) \tag{5.5}$$

It is well known that the solution to (5.4) is

$$\begin{aligned} x_1(t) &= x_{10}\cos\left(\sqrt{a}t\right) + \frac{x_{20}}{\sqrt{a}}\sin\left(\sqrt{a}t\right) \\ x_2(t) &= -\sqrt{a}x_{10}\cos\left(\sqrt{a}t\right) + x_{20}\sin\left(\sqrt{a}t\right) \end{aligned} \tag{5.6}$$

where $x_{10} = x_1(0)$ and $x_{20} = x_2(0)$ are the initial conditions. Plugging (5.6) into (5.5) and simplifying gives

$$I(x_1, x_2) = ax_{10}^2 + x_{20}^2 \tag{5.7}$$

Notice that I is independent of t and is only a function of the initial condition of the system. Another way of viewing the invariant is shown in Figure 5.1. Plotted in Figure 5.1(a), are trajectories of the system for various initial conditions. In Figure 5.1(b), the function $I(x_1, x_2)$ is plotted for each x_1 and x_2. As can be seen, I is constant along each trajectory.

Another characteristic related to Hamiltonian systems is volume-preserving flow. Recall that the flow of a dynamical system is given by its vector field $f(x)$. A flow is **volume preserving** if for any set of points in the state-space, the "volume" of those points does not change after they evolve through f. It is called "volume" as a general term even though it is technically a volume only when dealing with three-dimensional space. Figure 5.2 illustrates this concept in three dimensions. An initial set of points is shown as X_0, and these points occupy a certain volume in the state-space. After plugging each point of X_0 into f and evolving the trajectory for some time t, a different volume of the state-space is occupied, shown as X_t. If f is a volume-preserving flow, then the volume occupied by X_0 and X_t are the same for any value of t.

If we define the new set of points by $\varphi_t(X_0) = X_t$, then mathematically, volume preservation is expressed as

$$\int_{X_0} dx = \int_{\varphi_t(X_0)} dx \tag{5.8}$$

A flow is volume preserving if it is divergence free.[1] That is,

$$\operatorname{div} f = \frac{\partial f_1}{\partial x_1} + \frac{\partial f_2}{\partial x_2} + \cdots + \frac{\partial f_N}{\partial x_N} = 0 \tag{5.9}$$

[1] This is Liouville's theorem, whose proof is formulated in this notation in Wiggins (2003).

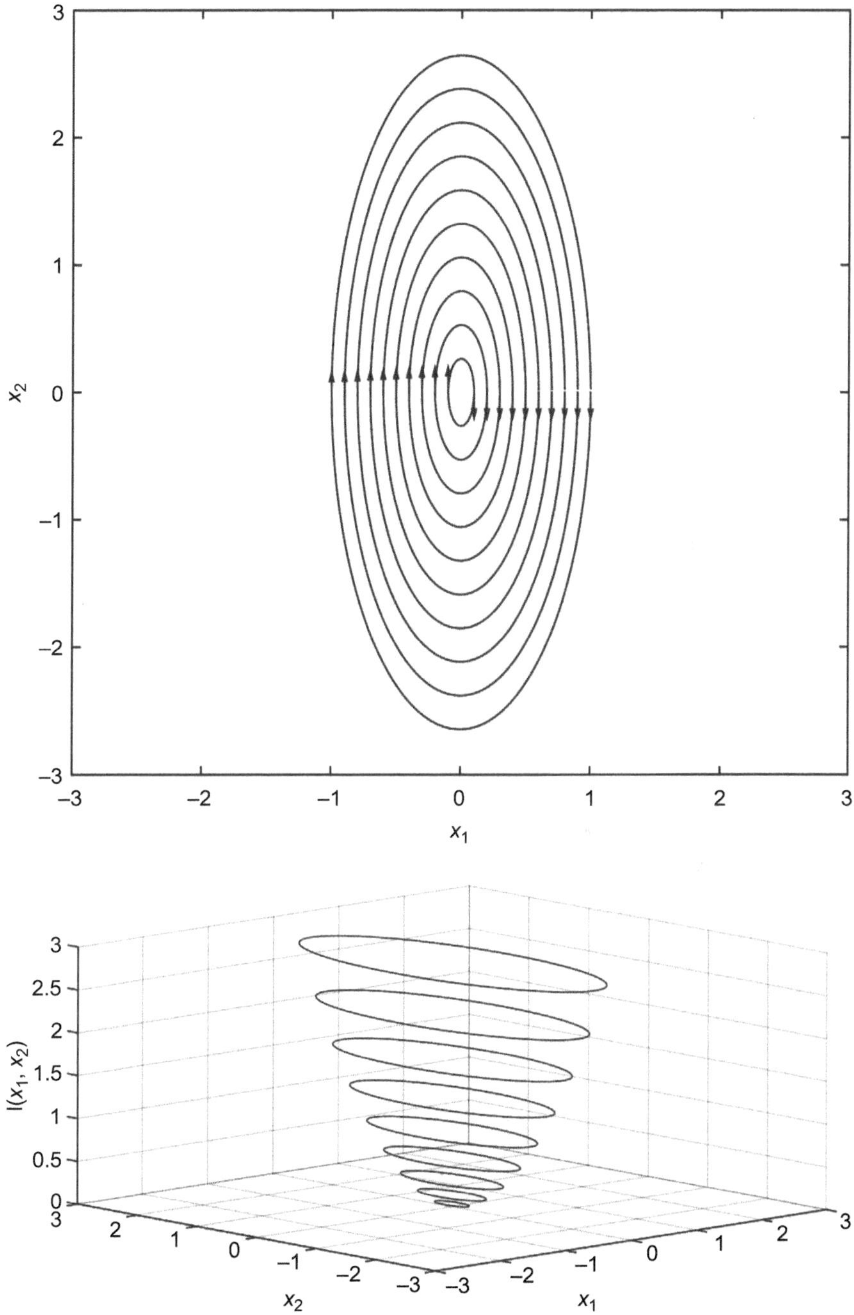

Figure 5.1 The phase plot of the harmonic oscillator (a) and the plot of $I(x_1, x_2)$ (b) along each trajectory of the phase plot.

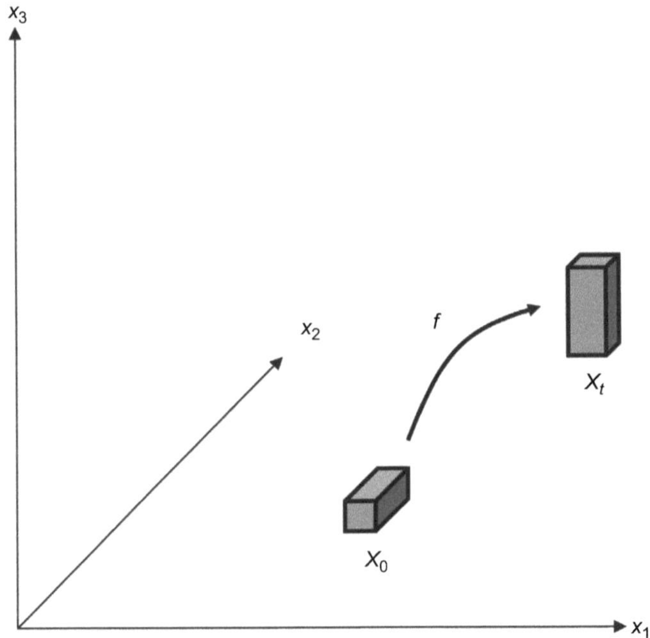

Figure 5.2 The flow f is volume preserving if the volumes of X_0 and X_t are equal.

Revisiting the harmonic oscillator example, the flow of (5.4) is

$$f = \begin{bmatrix} x_2 \\ -ax_1 \end{bmatrix} \tag{5.10}$$

Then the divergence of f is

$$\operatorname{div} f = \frac{\partial f_1}{\partial x_1} + \frac{\partial f_2}{\partial x_2} = 0 \tag{5.11}$$

and thus f for the harmonic oscillator is volume preserving. Figure 5.3 shows volume conservation with this system for $a = 7$. The initial set of points is the rectangle X_0. This set evolves to the parallelograms X_1 and $X_{1.5}$. The phase plot trajectories of the corner points of X_0 are shown to demonstrate how the initial rectangle moves and stretches. Because the flow of (5.4) is volume preserving, the areas of X_0, X_1, and $X_{1.5}$ are equal.

Hamiltonian systems first came about through a reinterpretation of Newton's second law applied to point particles in a conservative system. Later it was discovered that the theory applies to conservative systems with volume-preserving flows, but Hamiltonian systems have even more structure associated with them, as we will see in the next section.

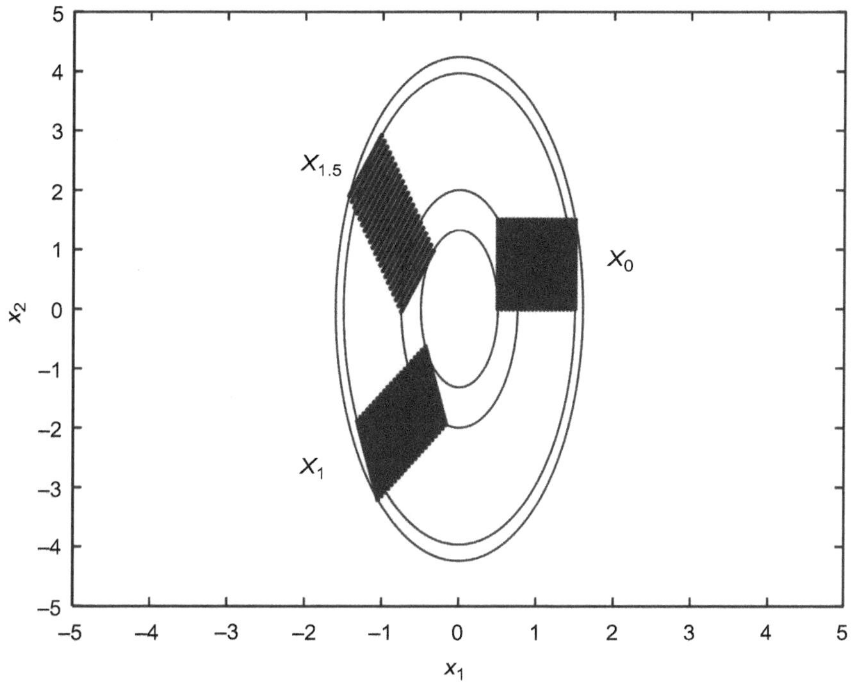

Figure 5.3 The volume-preserving characteristic of the flow in the undamped harmonic oscillator.

5.2 STRUCTURE OF HAMILTONIAN SYSTEMS

Following the idea of writing $F = ma$ for a system of N point particles, the formal definition of a Hamiltonian system is one whose dynamics satisfy Hamilton's equations:

$$\begin{aligned} \frac{dq_i}{dt} &= \frac{\partial H}{\partial p_i} \\ \frac{dp_i}{dt} &= -\frac{\partial H}{\partial q_i} \end{aligned} \tag{5.12}$$

where $i = 1, \ldots, N$. The variables p_i and q_i represent the momentum and position of mass m_i, respectively. The function H is known as the **Hamiltonian,** and it is an expression for the total (kinetic and potential) energy of the system given by

$$H(p, q) = \frac{1}{2}\sum_{i=1}^{N}\frac{p_i^2}{m_i} + V(q) \tag{5.13}$$

where $V(q)$ is the potential energy of the system.

These ideas were first applied to conservative mechanical systems. However, the idea can be generalized, and one can define the Hamiltonian H to be a smooth function that maps each (p, q) to a scalar quantity. Because $p = [p_1, \dots, p_N]$ and $q = [q_1, \dots, q_N]$, the Hamiltonian space has dimension $2N$. Today one can find Hamilton's equations formulated in the more general way as (for a second-order system)

$$\begin{aligned} \dot{x} &= \frac{\partial H}{\partial y} \\ \dot{y} &= -\frac{\partial H}{\partial x} \end{aligned} \tag{5.14}$$

where H: $\Re^2 \rightarrow \Re$ is a smooth function.

There are a few interesting results related to the structure of Hamiltonian systems.

- H is constant along trajectories of the system, and phase plots can be constructed without knowing solutions of the system.
- In systems for which H is independent of time, x^* is an equilibrium point if and only if it is a critical point of H. In other words, all partial derivatives of H evaluated at x^* are zero.
- Eigenvalues of linearized Hamiltonian systems appear in pairs. If λ is an eigenvalue, then $-\lambda$ is also an eigenvalue.

We illustrate these characteristics in the next section by revisiting some of the previous examples in the context of Hamiltonian systems.

5.3 EXAMPLES

5.3.1 Harmonic Oscillator

Consider again the harmonic oscillator introduced in Section 4.2 with its model given by

$$\begin{bmatrix} \dot{x} \\ \\ \dot{x}_2 \end{bmatrix} = \begin{bmatrix} 0 & 1 \\ -\dfrac{k}{m} & 0 \end{bmatrix} \begin{bmatrix} x_1 \\ \\ x_2 \end{bmatrix} \tag{5.15}$$

Define its Hamiltonian H to be

$$H(x_1, x_2) = \frac{1}{2}\left(\frac{k}{m}x_1^2 + x_2^2\right) \tag{5.16}$$

which satisfies the definition because

$$\frac{\partial H}{\partial x_1} = \frac{k}{m} x_1 = -\dot{x}_2$$
$$\frac{\partial H}{\partial x_2} = x_2 = \dot{x}_1 \tag{5.17}$$

Because we know that this H is constant along the system trajectories, the phase plot is simply created by plotting $\frac{1}{2}\left(\frac{k}{m}x_1^2 + x_2^2\right) = c$ for various values of $c > 0$. Figure 5.4 shows the resulting ellipses for $\frac{k}{m} = 7$ and $c = 1, \ldots, 10$.

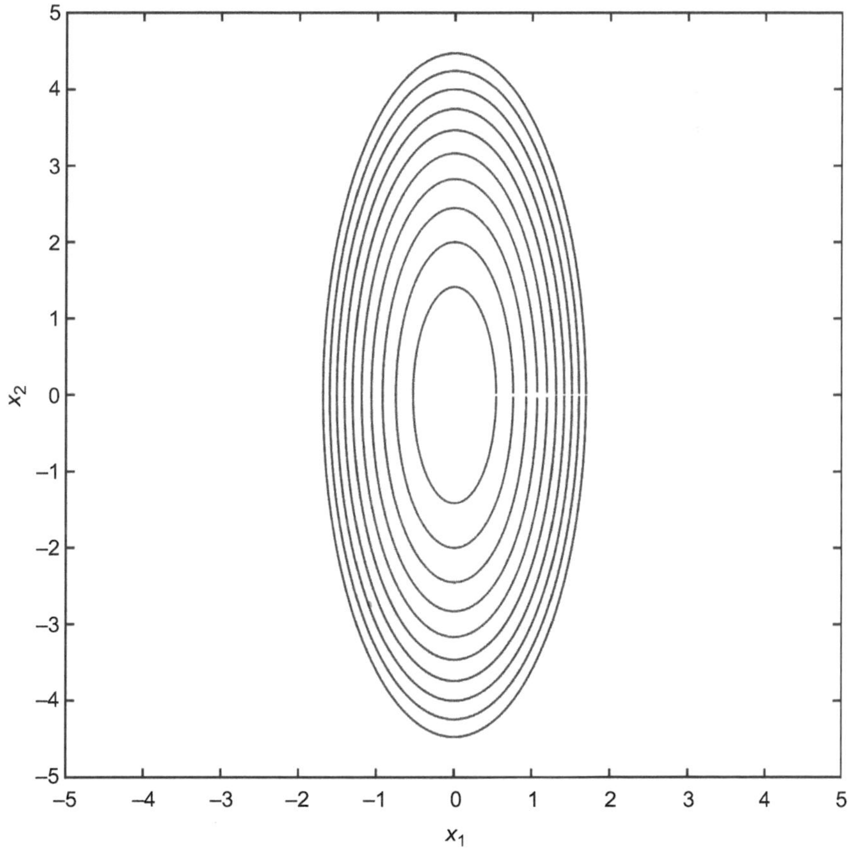

Figure 5.4 Trajectories of the harmonic oscillator obtained using the Hamiltonian. The direction of flow must be found from the vector field.

Although the curves for the trajectories can be obtained this way, the direction of travel along the trajectories is unknown. Direction information is easily obtained using the vector field.

To find the equilibrium points, we take partial derivatives of H and set them to zero.

$$\begin{aligned}\frac{\partial H}{\partial x_1} &= \frac{k}{m}x_1 = 0 \\ \frac{\partial H}{\partial x_2} &= x_2 = 0\end{aligned} \tag{5.18}$$

The harmonic oscillator therefore has one equilibrium point at (0, 0) as expected. The eigenvalues of the system are the eigenvalues of matrix $\begin{bmatrix} 0 & 1 \\ -\frac{k}{m} & 0 \end{bmatrix}$ and are located at $\lambda = \pm j\frac{k}{m}$.

If we introduce friction into the system, it is no longer Hamiltonian. The system equations are now

$$\begin{bmatrix} \dot{x} \\ \dot{x}_2 \end{bmatrix} = \begin{bmatrix} 0 & 1 \\ -\frac{k}{m} & -\frac{b}{m} \end{bmatrix} \begin{bmatrix} x_1 \\ x_2 \end{bmatrix} \tag{5.19}$$

The friction introduces energy dissipation into the system, so the system is no longer conservative. Also, the flow is no longer volume preserving. This is seen by mapping an initial set of points through the vector field for some time t. Figure 5.5 shows the sets of points X_0, X_1, and $X_{1.5}$ in the same way as Figure 5.3 but now with $b = 1$. The areas occupied by the sets of points are shrinking as t increases. Because the trajectories are all converging to the asymptotically stable equilibrium point at (0, 0), all initial conditions will eventually end up at (0, 0) and occupy the origin.

5.3.2 Pendulum

Let us revisit the pendulum system in (2.24) from the Hamiltonian point of view. For the undamped pendulum, the system equations are given as

$$\begin{aligned}\dot{x}_1 &= x_2 \\ \dot{x}_2 &= -\frac{g}{l}\sin x_1\end{aligned} \tag{5.20}$$

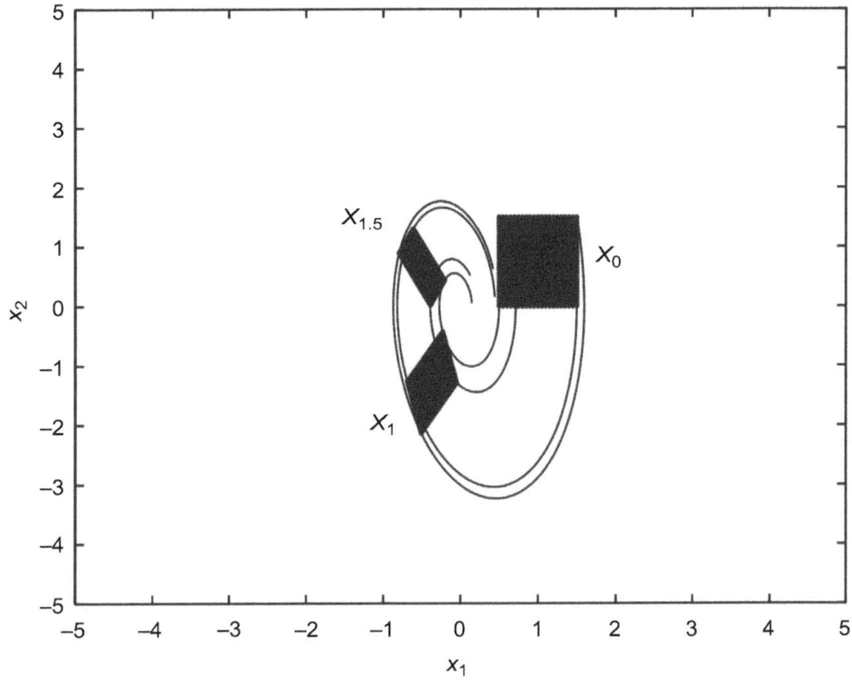

Figure 5.5 The flow of the damped harmonic oscillator is not volume preserving.

Define its Hamiltonian H to be

$$H(x_1, x_2) = 1 - \frac{g}{l}\cos x_1 + \frac{1}{2}x_2^2 \tag{5.21}$$

which satisfies the definition because

$$\begin{aligned} \frac{\partial H}{\partial x_1} &= \frac{g}{l}\sin x_1 = -\dot{x}_2 \\ \frac{\partial H}{\partial x_2} &= x_2 = \dot{x}_1 \end{aligned} \tag{5.22}$$

Notice that the Hamiltonian is not unique. It can always be modified by adding a constant without changing the basic properties. Next we generate the phase plot by generating curves $-\frac{g}{l}\cos x_1 + \frac{1}{2}x_2^2 = c$ for various values of $c > 0$. Figure 5.6 shows the result. Note that this phase plot looks a bit different from that shown in Figure 2.20 because the angular position is not restricted to be between $-\pi$ and π as it was in the previous phase plot example.

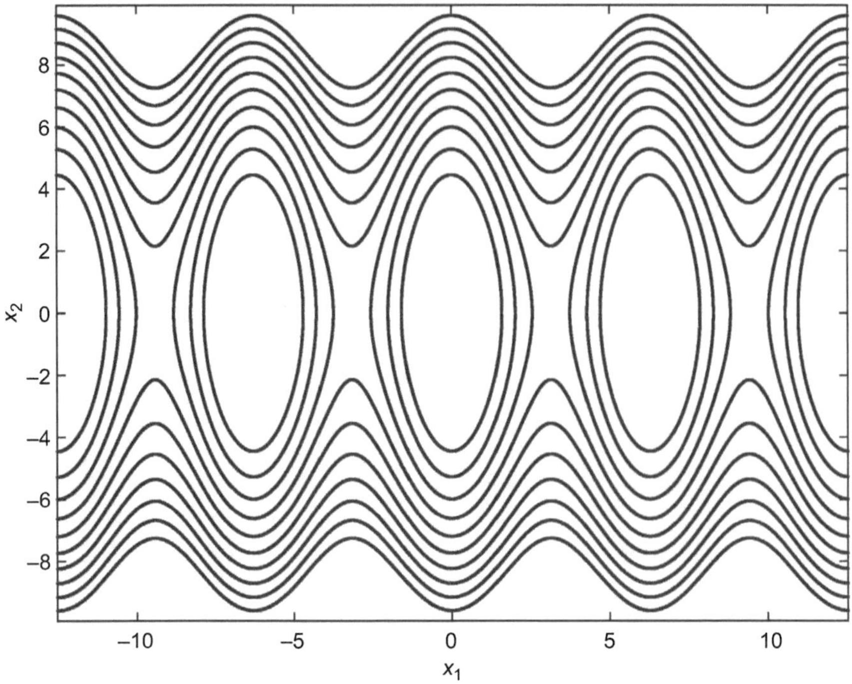

Figure 5.6 The phase plot of the undamped pendulum generated by the Hamiltonian.

The equilibrium points of the system are found by the relations

$$\begin{aligned}\frac{\partial H}{\partial x_1} &= -\frac{g}{l}\sin x_1 = 0\\ \frac{\partial H}{\partial x_2} &= x_2 = 0\end{aligned} \tag{5.23}$$

Thus, the equilibrium points for the pendulum are located at $(\pm n\pi, 0)$ as expected.

As with the harmonic oscillator, the damped pendulum whose model is

$$\begin{aligned}\dot{x}_1 &= x_2\\ \dot{x}_2 &= -\frac{g}{l}\sin x_1 - \frac{b}{ml^2}x_2\end{aligned} \tag{5.24}$$

is not a Hamiltonian system because energy is dissipated because of friction.

5.3.3 Population Dynamics

Recall that the dynamical equations for the predator—prey system are

$$\begin{aligned} \dot{x}_1 &= (a - bx_2)x_1 \\ \dot{x}_2 &= (-c + dx_1)x_2 \end{aligned} \tag{5.25}$$

where x_1 is the prey population and x_2 is the predator population. This system can be formulated as a Hamiltonian system by using a coordinate transformation. Define new variables y_1 and y_2 as

$$\begin{aligned} y_1 &= \ln x_1 \\ y_2 &= \ln x_2 \end{aligned} \tag{5.26}$$

Differentiating (5.26) gives

$$\begin{aligned} \dot{y}_1 &= \frac{1}{x_1}\dot{x}_1 \\ &= a - bx_2 \\ \dot{y}_2 &= \frac{1}{x_2}\dot{x}_2 \\ &= -c + dx_2 \end{aligned} \tag{5.27}$$

Solving (5.26) for x_1 and x_2 and substituting into (5.27) yields the new system equations

$$\begin{aligned} \dot{y}_1 &= a - be^{y_2} \\ \dot{y}_2 &= -c + de^{y_1} \end{aligned} \tag{5.28}$$

The Hamiltonian for the system can then be defined as

$$H(y_1, y_2) = cy_1 - de^{y_1} + ay_2 - be^{y_2} \tag{5.29}$$

5.3.4 Chaplygin Sleigh

We close this section with an example of a system that is *not* Hamiltonian. Typically, mechanical systems that are conservative are also volume preserving and Hamiltonian in nature. When dissipative forces, such as friction, are present, the system is not Hamiltonian. The Chaplygin sleigh is an example of a frictionless yet dissipative system.

The Chaplygin sleigh is shown in Figure 5.7. It consists of a rigid body moving in a plane on three contact points. The two front contact points are posts that slide along the frictionless surface with no constraints. The third

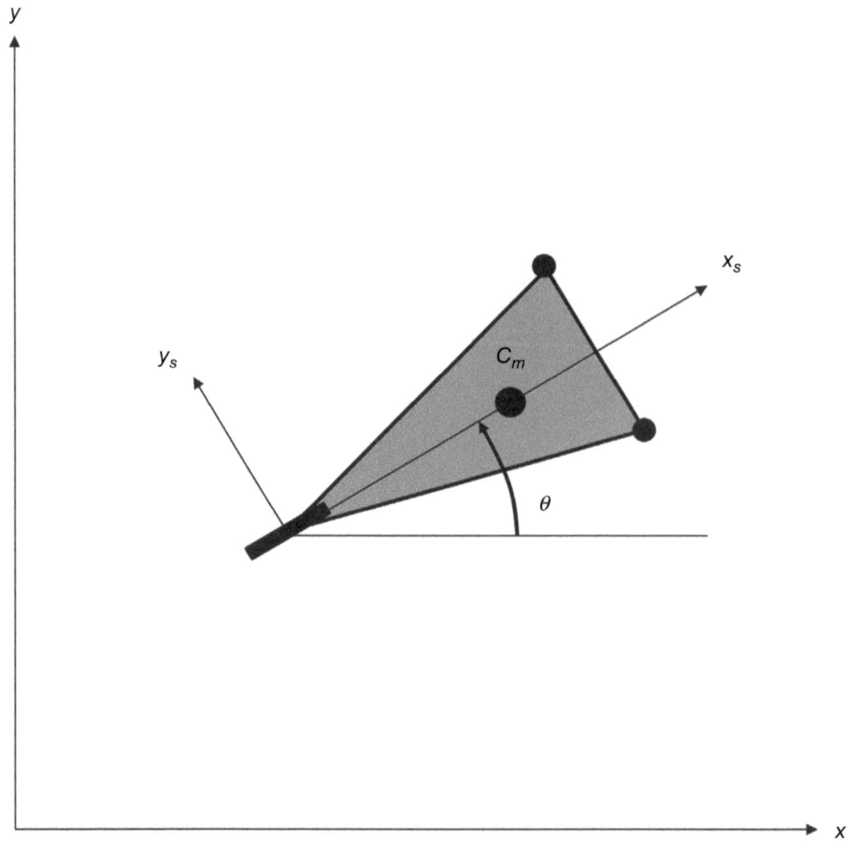

Figure 5.7 The Chaplygin sleigh is a rigid body moving in the plane with motion restricted to be along the x_s-axis.

point consists of a knife edge that also moves with no friction, but its movement is constrained to be along the edge, or the x_s-axis. The local frame of reference is attached to the body at the point where the knife edge contacts the surface. The center of mass is located at C_m, and d is the distance between C_m and the origin of the local frame.

If we define $v = \dot{x}_s$ and $\omega = \dot{\theta}$, the equations of motion for this system are[2]

$$\begin{aligned} \dot{v} &= d\omega^2 \\ \dot{\omega} &= \frac{md}{I + md^2} v\omega \end{aligned} \tag{5.30}$$

[2] A more detailed discussion of the system can be found in Bloch (2003).

where m is the mass of the body and I is the moment of inertia about C_m. Figure 5.8 shows the phase plot of the Chaplygin sleigh with state variables (v, ω). The trajectories form ellipses, which all converge to the positive v axis. Even though the system is conservative, it is not volume preserving as shown in Figure 5.9. Because the trajectories converge to a line, the volume of a given starting set of points gets smaller as time increases.

The Chaplygin sleigh is an example of a system that exhibits **nonholonomic constraints**. The word holonomic comes from the Greek words *όλος νόμος* meaning "entire" and "law," respectively. *Holonomic* was a term first used by Heinrich Hertz to mean "integrable." *Nonholonomic* therefore means "nonintegrable."

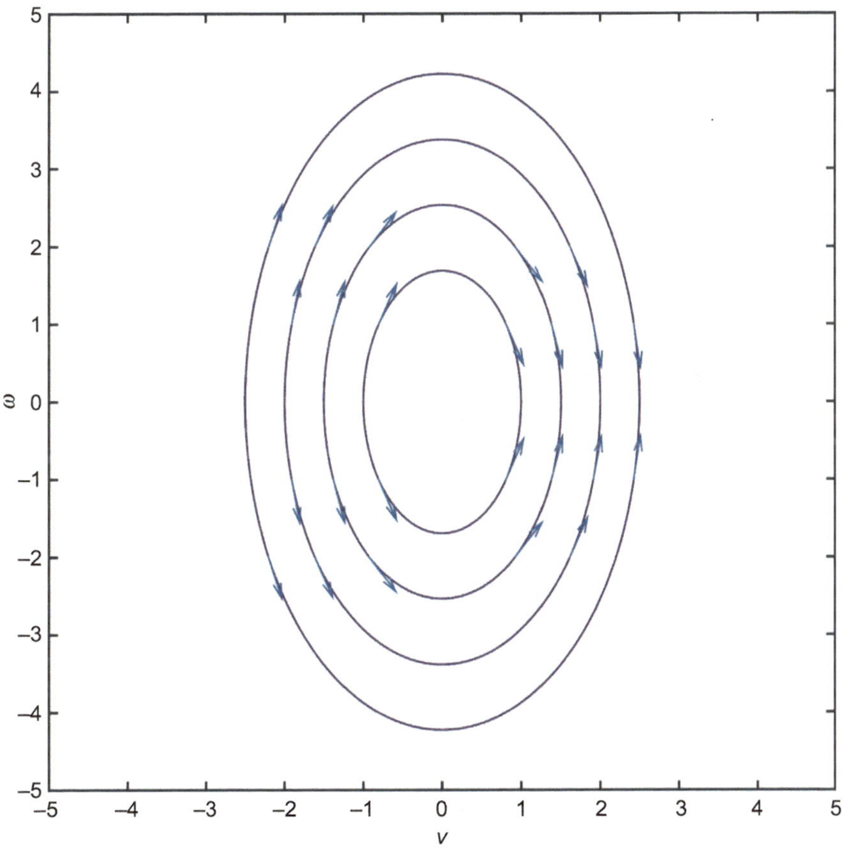

Figure 5.8 Phase plot of the Chaplygin sleigh.

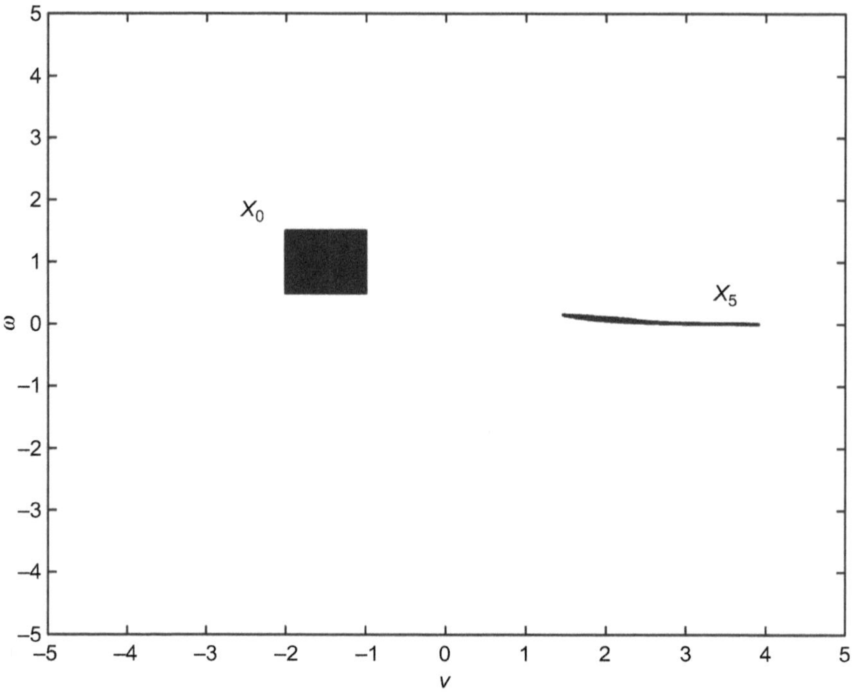

Figure 5.9 The flow of the Chaplygin sleigh is not volume preserving.

In mechanical systems, there are generally two kinds of constraints, geometric and kinematic. Geometric constraints are position restrictions, and kinematic constraints are velocity restrictions. If the kinematic constraints can be integrated to give geometric constraints, then it is considered a holonomic constraint. However, kinematic constraints may exist that do not restrict the geometry of the system, in which case they are known as nonholonomic constraints.

The knife edge of the Chaplygin sleigh imposes a nonholonomic constraint. The velocity is constrained because movement can only be in the direction of the edge and not perpendicular to it. However, despite this constraint, the body can move to any (x, y) position in the plane. It just needs to follow a trajectory in which it moves along its x_s-axis.

Nonholonomic systems are a large class of important dynamical systems. Readers are encouraged to investigate Bloch (2010), Bloch et al. (2005), and Neimark and Fufaev (1972) for further information.

5.4 CONCLUSION

This concludes the final chapter in our practical study of dynamical systems. We started with an overview of what dynamical systems are and why we study them. Then we moved into system modeling and how to represent models using various forms of mathematical expressions. From there we investigated important characteristics such as solutions, equilibrium points, and stability. In the final two chapters, we focused on nonlinear and Hamiltonian systems, respectively. It is hoped that readers have gained an appreciation for the practical side of dynamical systems through the application of the concepts to many different real-world examples with the mathematics worked through in detail. We have only touched the surface of the topics in dynamical systems, readers are encouraged to continue studying the many systems that are rich in interesting behaviors.

REFERENCES

Bloch, A. (2003). Dissipative dynamics in classical and quantum conservative systems. In J. Roenthal, & D. S. Gilliam (Eds.), *Mathematical Systems Theory in Biology, Communications, Computation, and Finance*. New York: Springer-Verlag.

Bloch, A. (2010). Nonholonomic mechanics, dissipation and quantization. In J. Levine, & P. Mullhaupt (Eds.), *Advances in the Theory of Control, Signals and Systems with Physical Modeling*. Berlin: Springer-Verlag.

Bloch, A., Marsden, J. E., & Zenkov, D. V. (March 2005). Nonholonomic Dynamics. *Notices of the AMS, 52*(3), 320–329.

Neimark, J., & Fufaev, N. A. (1972). *Dynamics of Nonholonomic Systems*. Providence, RI: American Mathematical Society.

Wiggins, S. (2003). *Introduction to Applied Nonlinear Dynamical Systems and Chaos* (2nd ed.). New York: Springer-Verlag.

FURTHER READING

Hirsch, M., Smale, S., & Devaney, R. L. (2013). *Differential Equations, Dynamical Systems, and an Introduction to Chaos* (3rd ed.). Waltham, MA: Academic Press.

Lynch, S. (2004). *Dynamical Systems with Applications Using MATLAB*. Boston: Birkhauser.

Meiss, J. (2007). *Differential Dynamical Systems*. Philadelphia: Society for Industrial and Applied Mathematics.

INDEX

Note: Page numbers followed by "f" indicate figures and "t" indicate tables.

FSC
www.fsc.org
MIX
Papier | Fördert gute Waldnutzung
FSC® C083411

Kolibri 360 GmbH
Ferdinand-Jühlke-Straße 7
99095 Erfurt, Deutschland
produktsicherheit@kolibri360.de